Safety & Security Engineering Series

Series Editor: S. Mambretti, *Politecnico di Milano, Italy*

Titles in Series

Flood Risk Assessment and Management
Landslides

Flood Risk Assessment and Management

Edited by

S. Mambretti
Politecnico di Milano, Italy

WIT*PRESS* Southampton, Boston

Safety & Security Engineering Series

Series Editor: S. Mambretti, *Politecnico di Milano, Italy*

SAFETY & SECURITY ENGINEERING

Editor: S. Mambretti, *Politecnico di Milano, Italy*

Published by

WIT Press
Ashurst Lodge, Ashurst, Southampton, SO40 7AA, UK
Tel: 44 (0) 238 029 3223; Fax: 44 (0) 238 029 2853
E-Mail: witpress@witpress.com
http://www.witpress.com

For USA, Canada and Mexico

WIT Press
25 Bridge Street, Billerica, MA 01821, USA
Tel: 978 667 5841; Fax: 978 667 7582
E-Mail: infousa@witpress.com
http://www.witpress.com

British Library Cataloguing-in-Publication Data

A Catalogue record for this book is available from the British Library

ISBN: 978-1-84564-646-2
eISBN: 978-1-84564-647-9
ISSN (print): 2047-7686
ISSN (online): 2047-7694

Library of Congress Catalog Card Number: 2011935060

The texts of the papers in this volume were set individually by the authors or under their supervision.

Printed in Great Britain by Polestar, Wheatons.

Preface

This volume is the first of a series the Publisher decided to launch to cover different aspects related to Safety and Security Engineering in order to reach a comprehensive view on risk mitigation.

The volume is devoted to floods, as one-third of the annual natural disasters and economic losses, and more than half of the respective victims are flood-related. A burgeoning global population and growing wealth, particularly in the last two or three decades, have increased the risk and the demand for protection from flooding. These features, together with climate change predictions and urban development, are affecting the way flood risk is managed.

Knowledge and scientific tools play a role of paramount importance in the strain of coping with flooding problems, along with the capacity building in the context of political and administrative framework. Therefore, governments need to establish clear institutional, financial and social mechanisms and processes for flood risk management in order to ensure the safety of people and property and, thereby, contribute to both flood defence and sustainable development.

The present volume contains selected papers presented at the Conferences organized by the Wessex Institute of Technology. The papers have been revised by the Authors, in order to be up-to-date and integrated in the book, which covers different topics:

- Risk assessment
- Mathematical models for flood propagation
- Effect of topographic data resolution
- Social and psychological aspects
- Decision making and management
- Legislations and directives

- Alternatives in flood protection
- Response and recovery
- Damages and economic – related problems
- Case studies

The Editor is very grateful to the Authors for their excellent contribution, and to the WIT Press for the availability in publishing a new Series in these very important subjects. The quality of the material makes the volume a most valuable and up-to-date tool for professionals, scientists, and managers to appreciate the state-of-the-art in this important field of knowledge.

The Editor
Milano, Italy
2012

Contents

Flood risk assessment in the Hornád and Bodrog river basins, Slovakia

M. Zeleňáková
Institute of Building and Environmental Engineering, Technical University of Košice, Slovakia

Abstract

The risk assessment process in general provides a way to develop, organize and present scientific information so that it is relevant to environmental decisions in the watershed. Floods have the potential to cause fatalities, displacement of people and damage to the environment. Flood risk means the combination of the probability of a flood event and of the potential adverse consequences for human health and the environment associated with a flood event. Flood risk assessment evaluates these potential adverse impacts that floods have on humans, the environment, cultural heritage and economic activity. The paper deals with flood risk assessment in general and in the Hornád and Bodrog river basin which is located in the eastern part of Slovak republic.

Keywords: flood risk, risk assessment, watershed, likelihood, adverse impacts, environmental risk

1 Introduction

Floods have the potential to cause fatalities, displacement of people and damage to the environment. Floods are natural phenomena which cannot be prevented. However, some human activities (such as increasing human settlements and economic assets in floodplains and the reduction of the natural water retention by land use) and climate change contribute to an increase in the likelihood and adverse impacts of flood events Directive 2007/60/EC [5].

Flood risk management is a process which comprises pre-flood prevention, risk mitigation measures and preparedness, backed up by flood management actions during and after an event. Flood risk management deals with the societal task of analysing, assessing and reducing flood risks considering all relevant physical, environmental and societal processes. It is essential that action will only be taken in areas where potential significant flood risks exist or are reasonably foreseeable in the future. If in a particular river basin, sub-basin or stretch of coastline no potential significant flood risk exists or is reasonably foreseeable in the future; EU Member States would be able to identify them in the preliminary flood risk assessment. For these river basins and sub-basins no further action would have to be taken [9].

The background for the process of flood risk assessment is environmental risk assessment. Environmental risk assessment evaluates the potential adverse effects that human activities have on the plants and animals that make up ecosystems. The risk assessment process provides a way to develop, organize and present scientific information so that it is relevant to environmental decisions. When conducted for a particular place such as a watershed, the environmental risk assessment process can be used to identify vulnerable and valued resources, prioritize data collection activity, and link human activities with their potential effects [15]. Environmental risk assessment is an objective, scientific process of identifying and evaluating the adverse risks associated with a hazardous substance, activity, lifestyle or natural phenomenon that may detrimentally affect the environment and human health [6]. When applied in a watershed context, risk assessment methods can help bring scientific data into environmental decisions [11]. The aims of assessment are to introduce a sound science-based assessment method to people working in watersheds; and to point out how using the methodology makes environmental assessment data more useful to managers.

2 Process of flood risk assessment

Flood risk means the combination of the probability of a flood event and of the potential adverse consequences for human health, the environment, cultural heritage and economic activity associated with a flood event 2007/60/EC [5].

Based on available or readily derivable information, such as records and studies on long term developments, in particular impacts of climate change on the occurrence of floods, a preliminary flood risk assessment shall be undertaken to provide an assessment of potential risks [11]. The assessment shall include a description of the floods which have occurred in the past and which had significant adverse impacts on human health, the environment, cultural heritage and economic activity and for which the likelihood of similar future events is still relevant, including their flood extent and conveyance routes and an assessment of the adverse impacts they have entailed.

According to 2007/60/EC [5] EU Member States shall complete the preliminary flood risk assessment by 22 December 2011. On the basis of this assessment they shall identify those areas for which they conclude that potential significant flood risks exist or might be considered likely to occur.

Above mentioned requests were taken into consideration to fulfil the aims of this paper. Background for the methodology of the flood risk assessment in the Hornád and Bodrog watershed (basin) is the general process of environmental risk assessment. The environmental risk assessment process consists of three main phases, seen in Figure 1: problem formulation, risk analysis and risk characterization [3, 8, 13, 14, 16]. Three additional components appear in the flow chart: planning, risk management, and iterative monitoring and data acquisition.

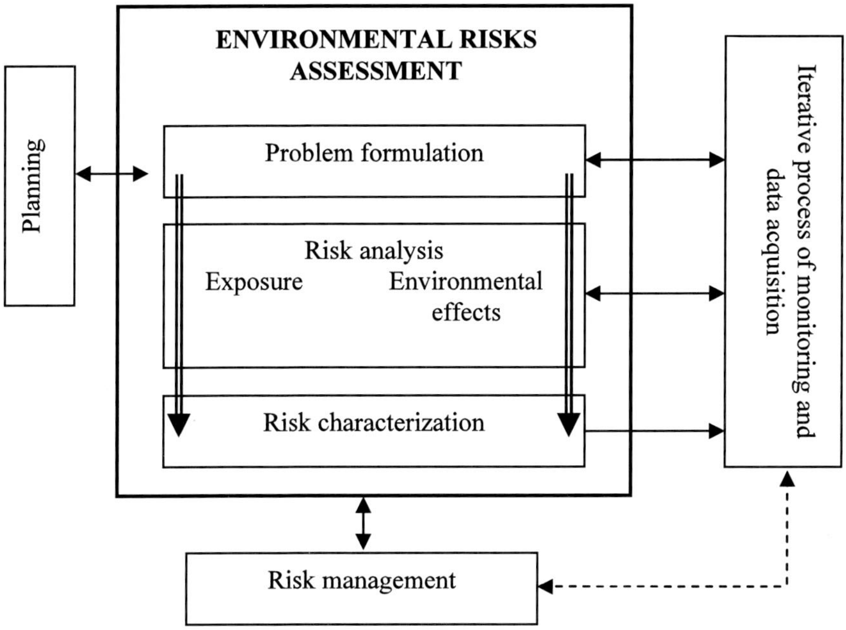

Figure 1: Flow chart of environmental risk assessment.

During planning, scientists and managers with input from stakeholders seek agreement on the focus, scope and complexity of an assessment. Then the formal risk assessment process commences with problem formulation during which key questions, conceptual models and an analysis plan are developed. The analysis phase evaluates the exposure of valued resources to stressors and the relationship between stressor levels and environmental effects. During risk characterization, the risks are described and if possible estimated quantitatively, forming the basis for the assessment's conclusions and a report. GIS tools are often used [10]. Monitoring and new data acquisition may occur in support of any of these phases, wherever needed. After completion, the risk assessment's findings are communicated to the managers, who determine a course of action.

3 Flood risk assessment in the Hornád and Bodrog watershed

The watershed, a hydrologically-bounded ecosystem, is a logical unit for environmental risk management. Watershed managers need a process for determining which environmental features in the watershed are at risk and choosing the best actions to protect them [2]. The process of risk assessment is accomplished by evaluating the likelihood that adverse environmental effects may occur as a result of exposure to stressors. The Hornád and Bodrog watershed (Figure 2) is situated in the eastern part of Slovakia.

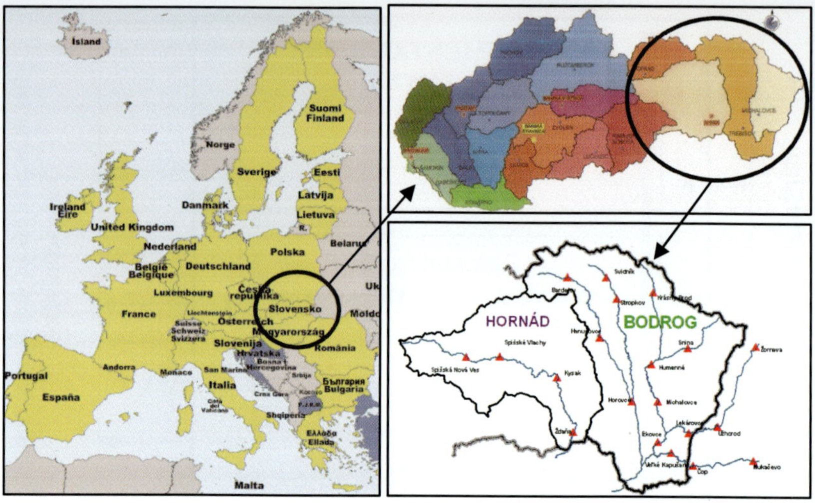

Figure 2: Hornád and Bodrog river basin.

The worst affected area by floods in Slovakia is the eastern part – Hornád and Bodrog river basins, particularly in the Topľa, Ondava, Torysa, and Laborec river basins in recent years [15]. Many floods have had tragic consequences and caused huge material damage.

3.1 Problem formulation phase

The problem formulation phase includes gathering available information about the watershed, its valuable resources potentially at risk, stressors and exposure opportunities, and environmental effects are a practical starting point.

Available information on valuable resources, stressors and effects is used to:

- Identify and select the specific subjects of the assessment (the assessment endpoints) – assessment endpoints are selected which provide a link between what can be measured (e.g. water level in stream) and one or more management objectives (e.g. ensuring flood protection). Act No. 7/2010 of Slovak republic on the flood protection [1] states three grades of flood according to the water level in the stream. The levels differ for the different river stations. Twenty river stations in the Hornád and Bodrog watersheds with the height in centimetres for the different flood grades (FG) were assessed. For the purposes of preliminary flood risk assessment daily data of water levels in the river stations for eight years period were assessed. The data were provided by Slovak Water Management Enterprise, s.c. Košice.

- Produce a conceptual model and associated questions that the assessment may address - the conceptual model describes, in narrative and graphical form, relationships among human activities, stressors, and the effects on valued resources [5]. At this point in the assessment these relationships are based on best professional judgement, but usually not yet quantified; yet the framework for analysis and assessment is clearly described therein. The simple example of the conceptual model is presented in Figure 3.
- Define a plan of action for the analysis phase and measurements that are needed - this analysis plan documents the exposure and effects relationships that will be quantified in the analysis phase, the data needed and measures to be used, and how risks will be described.

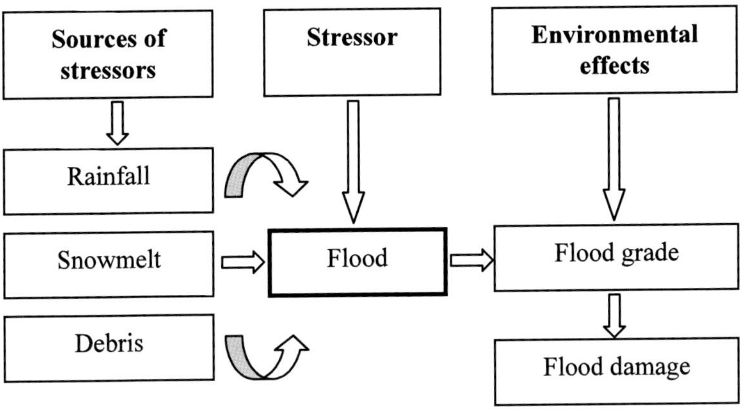

Figure 3: Conceptual model.

3.2 Risk analysis phase

Risk analysis phase focuses on the most important stressors, their exposure pathways, and the resulting environmental effects. The analysis phase includes characterisation of exposure – the manner in which a valuable resource contacts or co-occurs with a stressor – and characterisation of effects – the environmental response that occurs from exposure. The scope of the risk analysis may focus on the major stressor of concern or seek associations between stressor and impact. Flood is the major assessed stressor in this case. This phase analyses mainly:

- Exposure – analyzed by the temporal and spatial distribution of stressors in the environment. Maximal monthly heights of water level in each river station were evaluated in eleven-year period. The examples for river station Svidník is presented in Figure 4.
- Environmental effects – analyzed by describing stressor-response relationships. Stressor-response profile development should try to relate

the magnitude of the effect to the magnitude, duration, frequency, and timing of exposure. The number of flood situations in each river station was evaluated in period 2000-2010.

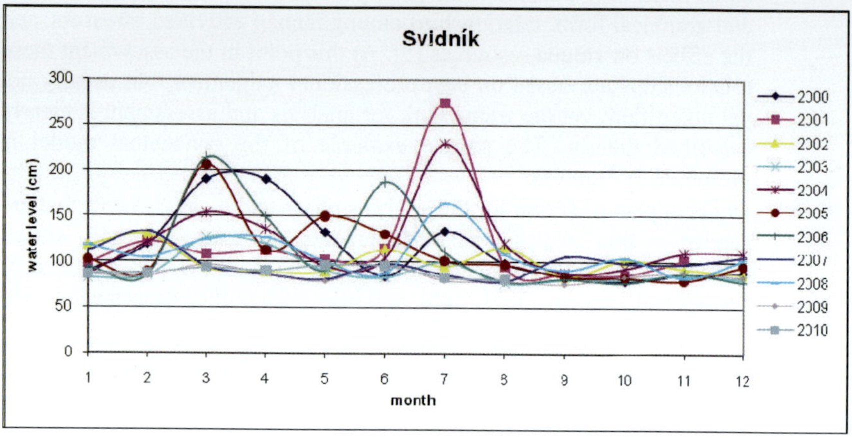

Figure 4: Heights of water level in Svidník river station.

3.3 Risk characterization phase

In the final phase of assessment – risk characterization phase – the likelihood and significance of adverse effects due to exposure to stressors are evaluated. The phase includes two major steps: risk estimation and risk description.

- Risk estimation, the first step, integrates the exposure profiles and the stressor-response profiles developed in the analysis phase. The likelihood and significance of adverse effects are evaluated according to the occurrence of flood grades (I. or II. or III.) in river stations [7]. Table 1 presents the average number (for the year) in the eleven-year period of flood grade occurrence in each river station. Sources of uncertainty may include measurement data, conditions of observation or human error.
- Risk description concludes the characterization phase with the preparation of an environmental risk summary and the interpretation of environmental significance. Interpreting environmental significance (Table 2) translates possible risk estimates into a discussion of their consequences for the watershed. Risk was counted from eqn. (1):

$$R_i = \sum_{i=1}^{n} \left(L_i \times C_i\right) = \left(L_1 \times C_1\right) + \left(L_2 + C_2\right) + \left(L_3 + C_3\right), \tag{1}$$

where: R – risk,

L_i – likelihood (occurrence of flood grade – I. FG = 1, II. FG = 2, III. FG = 3),

C_i – consequences (number of flood grades).

Table 1: Number of flood days at river stations.

Basin	River	River station	Number of I. FG	Number of II. FG	Number of III. FG
Hornád	Hornád	Ždaňa	309	22	5
		Kysak	146	15	5
		Sp.Vlachy	14	2	1
		S.N.Ves	0	0	0
Bodrog	Cirocha	Snina	0	0	0
	Laborec	Ižkovce	6	9	1
		Michalovce	11	1	0
		Humenné	7	2	0
		K. Brod	3	0	1
	Latorica	V.Kapušany	344	206	22
		Čop	230	345	189
		Mukačevo	0	0	0
	Ondava	Horovce	5	8	3
		Stropkov	6	9	2
		Svidnik	2	1	0
	Topľa	Hanušovce	14	11	5
		Bardejov	11	2	1
	Uh	Lekárovce	9	5	3
		Užhorod	0	0	0
		Žornava	8	1	2

Table 2: Risk acceptability and its significance.

Risk rate	Risk acceptability	Scale of risk	Significance of flood risk in basin
1	acceptable	1 – 10	Risks in basin are acceptable – current practice
2	moderate	11 – 20	Risks in basins are moderate – condition of monitoring
3	undesirable	21 – 30	Risks in basins are undesirable – flood protection
4	unacceptable	31 and more	Risks in basins are unacceptable – immediate flood protection

Table 3 represents the acceptability (risk rate) of flood risks for each of the Hornád and Bodrog river watersheds zones.

Table 3: Flood risk.

Basin	River	River station	Risk – quantitative assessment	Flood risk rate - acceptability
Hornád	Hornád	Ždaňa	41	4 - unacceptable
		Kysak	23	3 - undesirable
		Sp.Vlachy	2	1 - acceptable
		S.N.Ves	0	1 - acceptable
Bodrog	Cirocha	Snina	0	1 - acceptable
	Laborec	Ižkovce	3	1 - acceptable
		Michalovce	1	1 - acceptable
		Humenné	1	1 - acceptable
		K. Brod	1	1 - acceptable
	Latorica	V.Kapušany	91	4 - unacceptable
		Čop	165	4 - unacceptable
		Mukačevo	0	1 - acceptable
	Ondava	Horovce	3	1 - acceptable
		Stropkov	3	1 - acceptable
		Svidnik	1	1 - acceptable
	Topľa	Hanušovce	4	1 - acceptable
		Bardejov	1	1 - acceptable
	Uh	Lekárovce	3	1 - acceptable
		Užhorod	0	1 - acceptable
		Žornava	1	1 - acceptable

The results from preliminary flood risk assessment in the partial Bodrog and Hornád basins shows that unacceptable flood risks are in the lower parts of the river watershed where its tributaries flow into (Table 3).

Monitoring and data acquisition is iterative throughout all phases of the risk assessment process. Continued monitoring provides key feedback in that detection of continued adverse effects after risk management actions are in place indicates the need for more effective action.

A watershed management approach helps flood managers focus on the highest priority problems affecting surface waters in the basin. Preliminary flood risk assessment can be particularly useful in watersheds as a scientific method that includes steps for integration with planning, priority-setting, and decision-making.

Generally, watershed environmental risk assessment provides tools and information that may be used in managing risks. Risk management fills the crucial role of integrating the science-based assessment with the economic, social, legal, and political factors affecting management decisions and actions in the watershed [2].

4 Conclusion

Flood risk assessment is focusing on the likelihood of adverse effects of floods as a basic philosophy for making environmental decisions related to flood measurements in the watershed. Making good watershed management decisions requires science-based information that can be evaluated and priority-ranked in terms of the risks to the watershed. The results of preliminary flood risk assessment in the Hornád and Bodrog river basins is the identification of unacceptable flood risks in the lower part of the Hornád and Bodrog stream in the Eastern Slovak Lowland. This part of Slovakia is frequently exposed to floods and considering these streams is prior for flood measurements design.

The risk characterization should interpret the major risks and the environmental significance of the findings. Maps, simplified scoring systems, clearly defined evaluative criteria and limiting the numbers of stressors and effects addressed all help to assess effectively. Graphs are one of the best analytical tools for describing relationships between investigated attributes and impacts. Summary tables are an effective approach to display the most meaningful information in one condensed exhibit.

Other benefits of using this process in basins assessment are also evident. The improvements in the coordination associated with the planning can bring priority issues into focus and the increased awareness of basin flood problems and their relative priority can prompt other independent actions to improve flood control.

The paper presents application of flood risk assessment in condition of eastern part of Slovakia.

Acknowledgements

The Centre was supported by the Slovak Research and Development Agency under the contract No. SUSPP-0007-09.

Thanks to the support of APVV project SK-PL-0022-09.

References

[1] Act no. 7/2010 of Slovak republic of 2 December 2009 on the flood protection.

[2] Bendíková, M., *Environmental risk assessment of damage at water Environment.* Proc. of 3rd Int. Conf. of PhD Students: Engineering Sciences, Miskolc, p.17-23. 2001.

[3] Bendíková, M., *Environmental risks in condition of chosen water streams in Slovakia*. Thesis, Košice: TU, SvF, 2004.

[4] Bendíková, M., Švecová, A., *Sources of Pollution in Watersheds and Environmental Risks*. Zeszyty naukowe: Budownictvo i inžyniera środowiska 2. Rzeszów: Politechnika Rzeszowska, p.41-46. 2004.

[5] Directive 2007/60/EC of the European Parliament and of the Council of 23 October 2007 on the assessment and management of flood risks.

[6] Environmental Risk Assessment. http://europa.eu.int/comm/development /body/theme/environment/envman-149.html

[7] Gaňová, L., *Flood risk assessment and management*. Diploma work. Košice: TU, SvF, 2010.105 p.

[8] *Guidelines for Ecological Risk Assessment.* US EPA: Office of Research and Development, Risk Assessment Forum, Washington, DC. EPA/630/R-95/002f, 1998.

[9] Key Elements of the EU Directive. http://www.floodsite.net/html /eu_directive_keyelements.htm

[10] Kozáková, Ľ., Kuzevič, Š., Kuzevičová, Ž., Using geographic information systems for environmental analyses. *Metalurgija*. Vol. 45, No. 3, p.264. 2006.

[11] Šlezingr, M., Korytárová, J., Practical Using of Methodology of the Potential Flood Loss Assessment, In: *Wasserbaukolloquium 2007*, TU Dresden, Dresden, 2007

[12] Šlezingr, M., Reasons of the revitalization of watercourses. In: *Selected Scientific Papers*. Vol.4, Issue 1, p.69-80, 2009

[13] Watershed Ecological Risk Assessment. http://www.epa.gov/watertrain /ecorisk/index.html

[14] *Workshop Report on Characterization Ecological Risk at the Watershed Scale.* US EPA: Office of Research and Development. Washington, DC 20460. EPA 600-R-99-111, 2000.

[15] Zeleňáková, M., *Flood risk assessment*. Košice: TU, SvF, 101 p., 2009.

[16] Zeleňáková, M., Process of environmental risk assessment in watershed. *Transactions of the Universities of Košice*, No.1, p.18-26. 2006.

Mathematical models in flood management: overview, laboratory tests and case study

D. De Wrachien[1] & S. Mambretti[2]
[1]*Dept. of Agricultural Engineering, State University of Milan, Italy*
[2]*DIIAR, Politecnico di Milano, Italy*

Abstract

One-third of the annual natural disasters and economic losses, and more than half of the respective victims are flood related. These hazards are likely to become more frequent and more relevant in the future, due to the effects of increase in population, urbanization, land subsidence and the impacts of climate change.

Knowledge and advanced scientific tools play a role of paramount importance in the strain of coping with flooding problems. In this context, flood modelling represents the basis for effective flood mitigation.

The modelling approach aims to provide the best means for assessing and, subsequently, reducing the vulnerability of rural and urban flood prone areas.

By using models, an attempt is made to replace trial and error based strategies, as practised in the past, with more physically-based measures of flood management and control. Mathematical models are the best tools, nowadays available, for the design of efficient flood protection strategies and excellent supporters of decision-makers.

With reference to these issues, the paper provides a review and a general description of the main features of the models currently used in flood management along with the characteristics of the experimental data required for models' calibration.

Moreover, to highlight the effectiveness and the resilience of these tools, some laboratory tests and case studies are presented.

Keywords: flood modelling, flood frequency analysis, deterministic models, flood studies worldwide.

1 Introduction

Floods are among the most damaging of natural hazards, and are likely to become more frequent, more relevant and more damaging in the future due to the effects of increase in population, urbanization, land subsidence and, to a certain extent, the impacts of climate change.

In this context, flood modelling represents the basis for effective flood mitigation. They also are essential for flood risk assessment of both the current situation and feasible future scenarios. The modelling approach aims to provide the best tool for assessing and, subsequently, reducing the vulnerability of rural and high value urban flood prone areas as well as industrial zones.

Within this context, mathematical models are used for the following purposes (Van Duivendijk [7]):

- simulation of flood waves in rivers and their floodplains;
- assessment of the effectiveness of certain flood protection measures on extent of flooding and damages;
- evaluation of flood damages;
- design and construction of flood risk maps for zoning purposes;
- analysis of the effects of infrastructure and urban developments, as well as changes in land use, on flood;
- flood forecasting and warning;
- education, to increase communication and public awareness.

2 Types of models

Research work on flood dynamics has traditionally specialised in different mathematical models (De Wrachien *et al.* [6]). They can roughly be categorized into stochastic and deterministic models.

Stochastic models are based on flood frequency analysis, defined as the means by which flood discharge magnitude is related to the probability of its being equalled or exceeded in any year or to its frequency of recurrence or return period.

Deterministic models are, generally, based on physical properties of elements that feature or influence the phenomenon under investigation, such as the catchment characteristics, the channel geometry, and the rainfall-runoff process.

2.1 Stochastic models

Estimation of floods corresponding to specified return periods is essential for the design of flood protection measures, assessment of regions at risk of flooding, and the pertinent management of flooded areas.

Due to climatic variability, which drives flood events, stochastic modelling has become widely applied to estimate the magnitude of flood corresponding to a specified risk (Chow *et al.* [5]; Stedinger *et al.* [14]).

2.1.1 Flood frequency analysis

Flood frequency analysis was developed, in the first instance, in response to the need for information for the safe and economical design of engineering structures, either for the conveyance of flood flows (bridges, culverts, diversion channels, reservoirs, spillways) or for the protection or mitigation from flooding of land and property (embankments and walls). Moreover, this procedure is currently used for planning purposes and the design of land-use categories based on flood zoning with respect to vulnerability.

Frequency analysis is most commonly applied to peak discharges, instantaneous or averaged over a specified duration. Analysis is carried out on an observed historic record of river flow with the aim of assessing future probabilities of exceedance. It is also usually assumed that there will be not temporal change in the underlying statistics due to climate variability or to land use changes.

A large number of probability distributions and methods of application have been used for interpolation or extrapolation.

The two main approaches to select the flood series for fitting a stochastic model to the observed floods are based on the series for maxima annual flows (MAF) or partial duration series of floods (PDF). The MAF series selects the maximum flood event for each year, while the PDF series consists of all flood peaks above a specified magnitude.

The MAF approach ignores the fact that the highest flows in some years can be lower than flood events in other years, and hence it excludes significant high flood events in the parameter estimation process. The PDF procedure, on the other hand, considers all significant flood events in its parameter estimation process and its size can extend far beyond the available years of recorded flows. Recently, a new relation between the return period of the PDF and the MAF series was proposed (Mohssen [12]), based on the assumption that flood events are independent. The General Extreme Value (GEV) distributions, in addition to the Generalized Pareto (GP) distribution have been fitted and applied to analyse flood events.

2.2 Deterministic models

Deterministic modelling (flood routing) can be defined as a mathematical procedure for predicting the changing magnitude, speed and shape of a flow wave as a function of time (i.e. the flow hydrograph) at one or more points along a watercourse. The watercourse may be a river, stream, reservoir, estuary, canal, discharge ditch, or storm sewer. The flow hydrograph can result from precipitation runoff, reservoir releases, landslides into reservoirs, or tides. Flood routing may be classified as hydrologic (lumped), or hydraulic (distributed), or hybrid.

2.2.1 Hydrologic models

Hydrologic modelling involves the balancing of inflow, outflow and volume of storage through the use of continuity equation. A second relationship, the storage-discharge relation, is also required between outflows rate and storage in the system. This implies that water surface is level through the watercourse, usually a reservoir or lake. More complex relationships are to be sought for long and narrow reservoirs or open channels where storage is a function of both inflow and outflow. Both graphical and mathematical techniques for solving the continuity equation have been proposed.

The attractiveness of hydrologic modelling consists in its relative simplicity compared with hydraulic models. However, these procedures neglect backwater effects and are not accurate for rapidly rising hydrographs routed through mild to flat sloping rivers. They are also inaccurate for rapidly rising hydrographs in long reservoirs.

Hydrologic models can be categorized as (Fread [8]):

- level-pool types, used for reservoirs;
- storage types, used for rivers;
- linear systems types, which assume that the routing channel is composed of linear reservoirs connected by linear channels.

2.2.2 Hydraulic models

A good understanding of a complex flooding event can only be achieved by means of hydraulic models, because the flow rate, velocity, and depth vary in space along the watercourse and across a floodplain and/or a flood prone area. Estimation of these properties can be obtained by using the complete differential equations of 1D or 2D unsteady flow, known as the De Saint Venant (SV) or shallow water (SW) equations. Those equations allow the flow rate and water level to be computed as functions of space and time rather than time alone, as in the lumped flow routing methods. Distributed flow routing based on the complete SV or SW equations is known as hydrodynamic routing.

In some cases, the governing equations can be simplified to a 1D continuity equations and a uniform flow relationship, referred to as *kinematic wave routing*, which implies that the discharge can be computed as a simple function of depth alone. Uniform flow implies a balance between gravitational and frictional forces in the channel. This assumption can rarely be justified, especially on very flat slopes where effects of water surface cannot be ignored.

Cases where other terms in the momentum equation for hydraulic routing must be taken into account include:

- upstream movement of tide or storm surges;
- backwater effects from downstream reservoirs and tributary inflows;
- flood waves in channels of very flat slope;
- abrupt waves caused by sudden releases from reservoirs or dam failures.

Selection of a flow routing model for a particular application is influenced by the relative importance placed on the following factors:

- the model's suitability to answer the user's questions;
- the model's accuracy;
- the type and availability of required data;
- the complexity of the mathematical formulation.

Distributed flow routing models are useful tools for determining floodplain depths, required heights of structures (bridges or levees), inundation maps for dam-break contingency planning, transient waves created in reservoirs by gate or turbine changes, landslide-produced waves in reservoirs and unsteady flow in storm sewer systems. The real flow process in each of these applications varies in all three space dimensions. However, normally the spatial variation of the flow characteristics can be approximated as varying in only one space dimension – the direction along the flow channel. Thus the 1D equations of unsteady flow are widely used.

2.2.3 Hybrid models

Until recently, hydraulic models were not reckoned a practical alternative for flood routing because they were considered not economically viable to obtain cross section data over the reaches involved in flood routing.

Recent investigations (Hicks [9]) have revealed that hydraulic routing can be successfully used to determine discharge hydrographs in reaches where little channel geometry data are available, by approximating the model reach by a rectangular channel.

It was found that this "limited geometry" modelling approach – based on 1D SV equations – could accurately determine discharge hydrographs, making it an effective and suitable alternative to hydrologic flood routing. It was also found that this hybrid model offers the advantage of operationally combining the flood routing and the determination of flood levels (Blackburn and Hicks [4]). In addition, the use of a hydraulic model opens up the potential for modelling more dynamic flood events such as ice jam release surges, which cannot be handled by traditional hydrological modelling approaches.

2.3 Flood forecasting

In practical applications, flood forecasting involves two steps.

First, a flood routing model (usually hydrologic) is used to obtain the flood peak by routing flood events between streamflow gauging stations. This flood wave must then be put into a hydraulic model based on detailed channel geometry in order to forecast flood events at key-sites.

3 Data: requirements and problems

It is a demanding task to precisely define the input data needed for flood modelling. The great variety of models and the different purposes for which they are designed and proposed make it difficult to devise general rules feasible for all situations.

Within this context, the most important question to be addressed when defining the quantity and quality of data to be used as input of a model is the purpose pursued. Modelling a flooding event in a specific area, caused by the failure of a dam, and over a wide rural plain due to heavy rain, concerns different data and asks for different approaches. In the event of dam failure, precise information on topography and soil characteristics are much more necessary than data on earlier flooding, while in flood forecasting over a broad area, data of historic flooding are essential.

Hydrologic models, based on empirical storage-flow relations to approximate momentum effects, require only streamflow hydrographs as inputs. Hydraulic models require additional, physical and altimetric data describing the channel geometry and the floodplain morphology, in order to forecast flood levels at key sites.

To this end, topographic data, referring to longitudinal and cross sections of the river through its banks and dykes or outlining flood zones play a role of paramount importance. Longitudinal and cross sections alone do not suffice for investigations involving 2D models, which require the ground to be divided into cells interrelated by means of weirs, culverts, or other hydraulic structures.

The topographic data resolution strongly affects the flood model efficiency.

Sound Digital Elevation Models (DEMs) must provide an accurate description of micro-topography (e.g. levees, embankments, roads, buildings) to create a computational mesh in which all the elements that affect flow dynamics and flood propagations are included. Advances in models and in remote sensing techniques make possible to generate high resolution DEMs for whole watersheds. The latest developments in airborne laser scanning, makes it feasible to produce high quality digital surface models (DSMs) with accuracies less than ± 25 *cm* depending on the land cover, slope, flight parameters and environmental conditions (Sole *et al.* [13]).

Hydrological/hydraulic flow characteristics are fundamental in flood modelling along with boundary conditions on depths and discharges.

Either a specified discharge or water-surface time series can be used for both upstream and downstream flow conditions. The downstream boundary condition can also be a critical flow section such as the entrance to a waterfall or a steep reach.

The resistance to flow in a watercourse may be parameterized by the Manning, Chézy, Darcy friction coefficients which represent the effect of roughness elements of the channel bed and particles as well as losses mainly due to dynamic bed morphology and vegetation. Best results are obtained when the friction coefficients are adjusted (calibrated) to reproduce historical observation of stage and discharge.

A good understanding of flooding events can only be achieved by means of 1D or 2D hydrodynamic models. These models have to simulate the movement of a relatively shallow layer of water (1–4 *m* deep) over a vast, nearly flat area. As soon as the depth of the water layer drops down to 0.2–0.5 *m* it becomes difficult to simulate the flow. The reason lies in the behaviour of this phenomenon in large, nearly inaccessible (during flooding) environments, in which at any point the direction, the velocity and the depth of the current vary as a function of location, ground level and time. As a matter of fact, the flood wave moves through a floodplain in the course of a few hours/days and during this short period of time it is very difficult to make the observations (water level, direction and speed of current) required for the calibration of complex unsteady flow models.

4 Laboratory tests and case studies

To highlight the effectiveness and the resilience of the hydrodynamic simulation models, a number of laboratory tests and case studies of flood mitigation and hazard assessment are featured. The first two deal with laboratory measurements reproducing dam-break waves, the last one concerns a river basin in Italy. After describing the methodologies, the chapter focuses on both the results achieved and the reliability of the procedures applied.

4.1 Calibration of 1D dam-break waves (laboratory tests)

Flash floods triggered by a sudden collapse of a dam (dam-break) are often characterized by the formation on shock waves with floating debris caused by many factors, such as valley contractions, irregular bed slopes and non-zero tail water depth. It is commonly accepted that a mathematical description of these phenomena can be accomplished by means of 1D or 2D SV equations (Ballos and Sakkas [2]; Bechteler *et al.* [3]; Aureli *et al.* [1]).

To validate the model, comparisons have been made between its predictions and experimental results carried out in the Hydraulic Laboratory of the Politecnico di Milano. The tests were performed with flows of both clear water and mixtures in a uniform geometry flume reproducing dam-break waves (Larcan *et al.* [10]; Mambretti *et al.* [11]).

To take into account different behaviours of the flow in the case of mixtures, the experimental data have been compared with the predictions of three rheological laws included in the one phase model (called "Water", "Fix Bagnold" and "Mobile Bagnold") and with those of the two phase model.

Comparisons show good agreement on the general shape that includes a steep front immediately followed by the maximum wave height and a decrease in flow depth down to an asymptotic value reached at the stoppage (figure 1).

The model can easily be extended to channels with arbitrary cross sections for both mature (non-stratified) and immature (stratified) debris flow routing, as well as for solving different problems of unsteady flow in open channels by incorporating the appropriate initial and boundary conditions.

4.2 Calibration of 2D dam-break waves (laboratory tests)

The 2D mathematical model is based on SV equations integrated with both the numerical schemes of Lax-Friedrichs and MacCormack. Tests have been carried out in the Hydraulic Laboratory of the Politecnico di Milano both with clear water and mixtures, having a channel 3.1 metres long and 0.5 metres wide which expands downstream with a steel plate to simulate the alluvial fan. Both the slope of the channel and of the plate are adjustable.

Tests have been recorded with a digital camera with resolution equal to 2 Megapixel and velocity equal to 25 frames per second. From the recorded movies, frames have been extracted and rectified in order to reduce the distortions. Then, the recorded waves have been compared with those simulated by means of the SV equations.

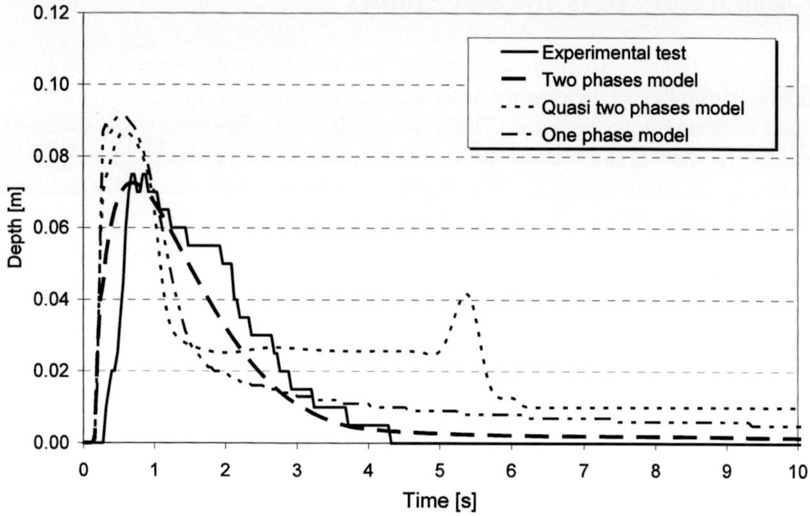

Figure 1: Debris flow wave in some characteristic sections of the experimental channel. Comparison between mathematical model and experimental results. Water-gravel, abs 140, conc. 40%, slope 20°, smooth bottom.

Results in clear water show significant agreement between recorded and simulated waves, as can be seen in the comparison reported in table 1 which gives the evolution of a wave on a smooth plate, for different volumes of water. An example of a picture of the experimental wave and its homologous computed by the mathematical model (different colours represent the different depths of the wave) is reported in figure 2. It is to be highlighted as no calibration has been performed on the model, in order to fit the experimental data, but the raw results have been used for comparison and shown.

The simulated wave has been computed using the Lax-Friedrichs numerical scheme, which is a first-order approximation and therefore less precise than McCormack's; however, it is much more stable and fast. Generally speaking, the simulated wave tends to underestimate the velocity of the experimental wave, but at the beginning of the transient flow, probably because of the numerical diffusion of the scheme. Moreover, higher volumes perform better, because of higher depths on the plate. All in all, the numerical methods seem to give a good approximation of the experimental wave, with errors that are common and acceptable in engineering practice.

4.3 River Lambro case study

In Milano the flood prone area of the Lambro River has been studied. The sector investigated was the area of the Lambro River within the Milano Municipality, which can be flooded in the case of an event with a 500-year return period.

Table 1: Comparison between experimental and simulated locations of the front wave on the plate simulating the alluvial fan.

Volume of water wave	Time $[s]$	Position of the front of the wave Experimental $[m]$	Numerical $[m]$	Error [%]
3 *litres*	0.40	0.8	0.8	0
	0.84	1.3	1.1	-15
	1.28	1.7	1.3	-23
	1.72	2.0	1.5	-25
6 *litres*	0.28	0.6	0.8	33
	0.60	1.2	1.1	-8
	0.92	1.6	1.4	-12
	1.24	2.0	1.7	-15
9 *litres*	0.24	0.6	0.8	33
	0.52	1.1	1.1	0
	0.80	1.5	1.4	-7
	1.08	1.9	1.7	-10

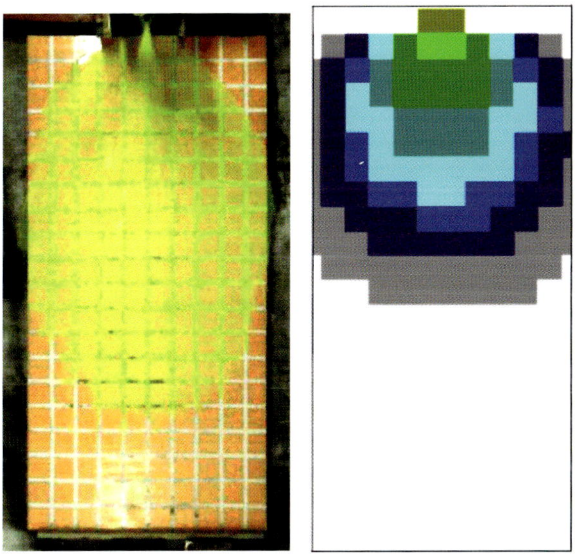

Figure 2: Experimental *Vs* numerical wave; $t = 0.80$ s, experimental wave $L = 1.5$ m; numerical wave $L = 1.4$ m.

A 2D model, based on the SV equations, has been built and calibrated using recordings of a large flood happened in 1951.

The model used for the simulations is an earlier version of the FLO-2D, with rectangular grid. The area, of about 20 km^2, has been divided in squared cells of

50 *m* x 50 *m*, each characterised by the ground elevation and the Manning roughness coefficient. Simulations have been carried out with three different discharge values: one for the incipient flood, one for a flood of 200 year return period, and the last with a 500 year return period discharge.

The model gives the depth and velocity for each cell, function of the time. Maximum values of depth and velocity have been computed for each cell and to each cell a degree of hazard has been assigned.

Figure 3 compares the reconstruction of the flooded areas in 1951 and the simulation carried out with the FLO-2D model. As can be seen, the flood prone area is very similar to that surveyed in 1951; however, in some cases the simulated flood seems to be even more reliable than the surveyed, as in 1951 they recorded flooded areas which are not connected to the river and therefore, if they were actually flooded, the reason was not to be attributed to Lambro.

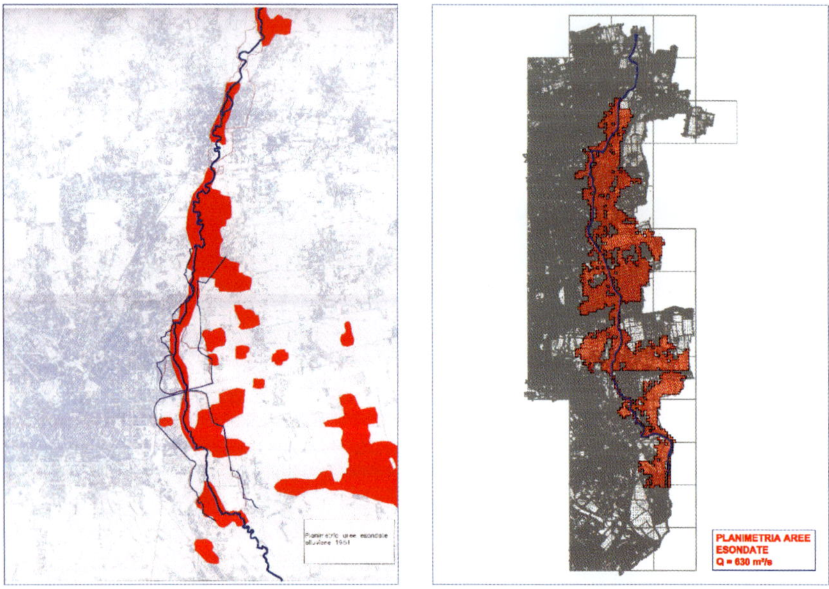

Figure 3: Reconstruction (*left*) and simulated (*right*) flooded area for the event of 1951 in Milano, due to the river Lambro.

5 Concluding remarks

Floods are among the most damaging of natural hazards, and are likely to become more frequent, more relevant and more damaging in the future due to the effects of increase in population, urbanization, land subsidence and, to certain extent, the impact of climatic change.

A flood event is a natural event of great complexity. The hydrological parameters of a flood (magnitude, frequency, celerity, volume, duration) reflect the stochastic behaviour of precipitation, interception, infiltration, evapotranspiration, soil moisture, overland and ground flows and river channel hydraulics. Models not only help in understanding these flood phenomena, but are also essential for flood risk assessment of the current situation and for assessment of expected changes.

By using models an attempt is made to replace trial and error-based strategies as practised in the past with more physically-based measures. The modelling tool aims to provide the best concept for assessing and reducing the vulnerability of rural and high-value urban flood-prone areas as well as industrial zones.

Within this context, the paper provides a review and a general description of the main features and characteristics of the mathematical models currently used in flood management, mitigation and control, along with the quantity and quality of experimental data to be used both as input and for model calibration.

Mathematical models are useful tools for the design of efficient flood protection strategies and excellent supporters of decision makers, since they enable a complex evaluation of cost benefit analysis of particular proposal for flood protection measures.

References

[1] Aureli F., Mignosa P., Tomirotti M. "Numerical simulation and experimental verification of dam-break flow with shocks" *Journal of Hydraulic Research*, 38, pp. 197-216, 2000

[2] Bechteler W., Kulish H., Nujic M. "2D dam-break flooding wave: comparison between experimental and calculated results" *Floods and Flood Management*, Ed. Saul, Dodrecht, 1992

[3] Bellos V., Sakkas J.G. "1D dam-break flood propagation on dry bed" *Journal of Hydraulic Engineering*, ASCE 113(12), pp. 1510-1524, 1987

[4] Blackburn J., Hicks, F.E. *Suitability of Dynamic Modelling for Flood Forecasting During Ice Jam Release Surge Events*, American Society of Civil Engineering: *Journal of Cold Regions Engineering*, 17(1) pp 18-36, 2002

[5] Chow V.T., Maidment D.R., Mays L.W. *Applied Hydrology* McGraw Hill, New York, 1988

[6] De Wrachien D., Mambretti S., Sole A. *Mathematical models in flood management: overview and challenges* Second International Conference on Flood Recovery Innovation and Response, pp. 61-72, 26 – 28 May 2010, Milan, Italy

[7] Duivendijk J. van *Manual on Planning of Structural Approaches to Flood Management.* ICID Guideline, New Delhi, India, 2005

[8] Fread D.L. "Flow Routing", Chapter 10, *Handbook of Hydrology* Maidment D.A. (Ed), McGraw Hill, New York, 1992

[9] Hicks F.E. "Hydraulic flood routing with minimal channel data: Peace River, Canada" *Canadian Journal of Civil Engineering*, Vol. 23, pp. 524-535, 1996

[10] Larcan E., Mambretti S., Pulecchi M., "A procedure for the evaluation of debris flow stratification." Proceedings of the 1st International Conference on *Monitoring, Simulation, Prevention and Remediation of Dense and Debris flow* Lorenzini, Brebbia and Emmauouloudis (eds), Rhodes, Greece, 7-9 June 2006

[11] Mambretti S., Larcan E., De Wrachien D. *1D modelling of dam – break surges with floating debris* Biosystems Engineering, Vol. 100(2), June 2008, pp. 297-308

[12] Mohssen M. "An insight into flood frequency for design floods." Proc. of the 1st International Conference on *Flood Recovery Innovation and Response*, WIT Press, London, 2008

[13] Sole A., Giosa L., Nolè L., Medina V., Bateman A. Flood risk modelling with LiDAR technology Proc. of the 1st International Conference on *Flood Recovery Innovation and Response*, WIT Press, London, 2008

[14] Stedinger J.R, Vogel R.M., Foufoula-Georgiou E. "Frequency Analysis of Extreme Events." Chapter 18, *Handbook of Hydrology* Maidment D.A. (Ed), McGraw Hill, New York, 1992

Effects of topographic data resolution and spatial model resolution on hydraulic and hydro-morphological models for flood risk assessment

E. Frank[1], G. Sofia[2] & S. Fattorelli[1,2]
[1]*Beta Studio srl, Italy*
[2]*Padova University, Italy*

Abstract

Uncertainties in topographic data have a significant influence on hydro-morphological and hydraulic predictions and therefore on flood risk assessment. In this work, the effects of topographic data resolution on the results of hydro-morphologic and hydraulic simulations are analysed using respectively the morphological bi-dimensional curvilinear model MIKE 21C and the 1D-2D hydraulic model Sobek. The studies have been carried out in two areas in Northern Italy: the river Torre and an alluvial plain on Adige river right bank. For both models, high-resolution LiDAR (Light Detection And Ranging) data were used to create the basic topographic information. An additional source for the topograpy of the Torre River has been provided by river cross-section data. Digital elevation models at different resolution have been created to test the effects of different grid cell sizes on the simulations. The impact of topographic information on hydraulic and morphological model results was evaluated for both areas through a comparison of results. Additionally, morphologic variations and position of erosional and depositional zones along the watercourse and variations in flood extent and dynamics have been investigated. The obtained results emphasise the importance of quality of input information for reliable results of hydraulic and morphological and sediment transport models. Criteria for selecting the optimal DTM resolution are suggested, based on the quality of available data.

Keywords: topographical resolution, LiDAR, hydro-morphological model, hydraulic model, sediment transport.

1 Introduction

Accurate representation of topography and appropriate resolution of data used to perform simulations are of prime importance in hydraulic and morphological modelling. Although, the first point has been studied by many researchers, [1–6], the second has been scantily treated.

There is often a dilemma in selecting the resolution of Digital Terrain Model (DTM): coarse resolution DTMs usually result in a lower topographical accuracy, but they require a smaller computational effort, while high-resolution ones better represent reality, but they lead to an excessive computational time. Modern technologies offer nowadays a large number of solutions to obtain and process high quality data and the development in airborne remote sensing data capturing techniques (LiDAR-Light Detection ad Ranging) offers a new support to the use of high-resolution data. Therefore, in practical cases, it is important to evaluate the effects of input data resolution on modelling simulations.

The aim of this paper is to study the effect of topographic data (LiDAR, cross-section) and spatial model resolution on hydraulic and hydro-morphological variables simulated with the mono-bidimensional model SOBEK and the bi-dimensional model MIKE 21C. This study indicates that different resolutions on topographic data produce important dissimilarities in simulation results, concerning morphological variations (distribution of sediment transport patterns, erosion rates) and flow hydraulics (flow mitigation, propagation time, flood extents). Comparing model results obtained with different grid dimensions, from higher to lower resolution, important details are lost, mainly as a consequence of averaging and internal interpolations of elevation data, which influences model performance and the reliability of simulation results.

2 Description of the models

In this study, the hydro-morphological variables have been simulated with the bi-dimensional model MIKE 21C developed by DHI [7], while flood simulations have been made through the integrated 1D-2D raster based Sobek flow modelling system developed by WL|Delft Hydraulics.

The modelling of flows within the channel, in both cases, is based on equations of continuity and conservation of momentum (De Saint Venant equations). These equations are solved by implicit difference techniques, with variables defined on a space-staggered computational grid.

In MIKE 21C the hydrodynamic solution is used to evaluate solid transport and the continuity equation of the sediment. The model updates at each time step the curvilinear grid and it represents erosion and deposition processes dynamically, considering a feedback between sediment transport and hydrodynamic conditions.

In Sobek, to represent 2D flow fields, water is transferred to the adjacent floodplains when bank full depth is exceeded, and flow is described in the 2D domain in terms of continuity and mass flux equations (Wright *et al.* [8]).

Both models require topographic information and boundary and initial conditions. Additionally, MIKE 21C requires sediment characteristics of bed (mean diameter, number of fractions, layer thickness and sediment fraction maps), and model coefficients (resistance, helical flow, transverse slope, etc).

2.2 Study areas

The hydro-morphological simulations have been carried out for the river Torre (A in fig. 1), while the flood modelling has been performed in an alluvial plain placed on the Adige river right bank (B in fig. 1).

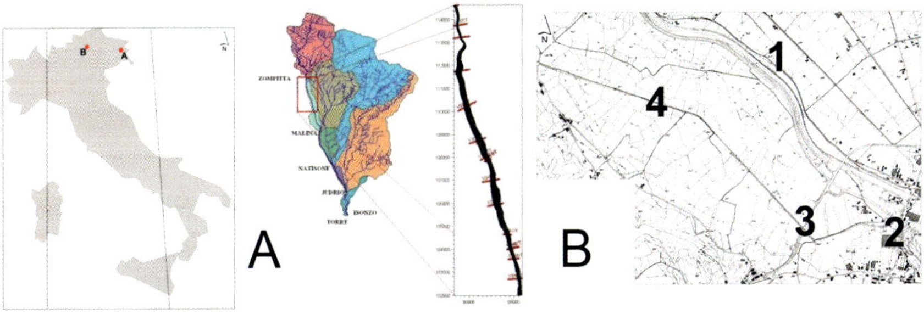

Figure 1: Location of the study areas: Reach of the river Torre (A) studied with the MIKE 21C simulations and Adige floodplain (B) studied with the SOBEK model (1. Adige River; 2. pumping station; 3. road embankment; 4. drainage channel).

The river Torre rises in the Italian Dolomites (north-east Italy) and flows for 66 km until it ends at the confluence with the river Isonzo. Its braided riverbed is formed of central islands partially covered with vegetation and extended floodplain areas, often used for agriculture. The portion of Torre analysed in this work is 12 km long (slope, S = 0.0046) and it is characterised by peak discharge of 670 m^3/s (T_R =100 years) for a precipitation of 12 hours (BETA Studio srl [9]). The riverbed is mainly composed of gravel, with a mean grain size (D_{50}) varying between 90 and 10 mm, which dominates the sediment movement. The reach is crossed by several structures (bridges, dams) that influence its hydraulic and morphological behaviour.

The flood modelling study area is an alluvial plain of about 8.3 km^2 crossed by a road embankment. In case of a bank break or a bank overflowing, water from Adige river would flow across the plain in the south-southeast direction and the road embankment (no. 3 in fig.1-B) would act as a barrier influencing the flow path.

2.3 Input data of the models

For both study areas, high-resolution LiDAR data with point density varying between 0.3–6 point/m^2, have been used to correctly represent topography. To model bathymetry within the MIKE 21C, an additional data source model has been provided by field surveys of 15 river cross-sections (numbered from 58 to 43), separated by approximately 900 m, with an average width of 170 m. DTM derived from cross sections was obtained by joining the principal points (upper-down banks on the right and left sides and bottom) through five 3D-lines drawn according to orthophotos and the Regional Technical Map.

2.3.1 Input data for hydro-morphological simulations

A curvilinear grid of the riverbed was constructed, considering floodplains and following the streamline directions, stream banks, levees, hydraulic structures, etc. Five computational grids with different resolution were constructed (fig. 2). The main characteristics are listed in table 1.

Two digital terrain models (2 m x 2 m grid) were generated from the basic topographical information, the first from LiDAR data (bathymetry 'a') and the second from cross-sectional data (bathymetry 'b'). These constituted the input data for the automatic generation of the curvilinear grid bathymetry in the MIKE 21C model.

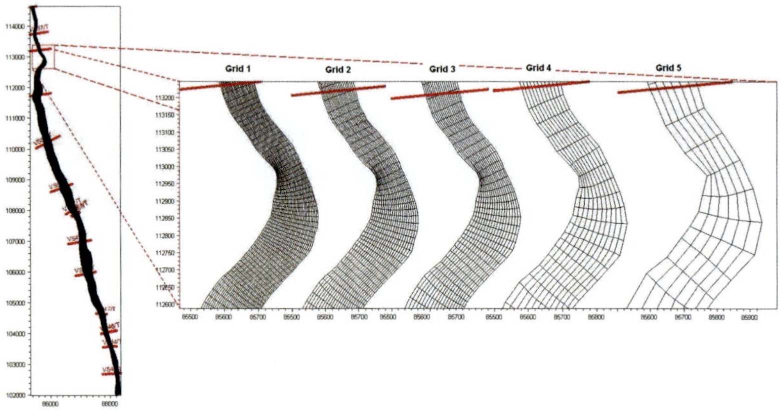

Figure 2: Curvilinear grids constructed with different cell dimensions. The grid order is shown according to refinement.

Table 1: Characteristics of computed grids. Average cell dimensions in cross-sectional direction (*dn*), in motion direction (*ds*), and number of cells is reported for each grid.

Grid number	Average dimensions			Number of grid cells		
	dn (m)	*ds* (m)	*dA*, area (m^2)	Cross-sectional direction	Motion direction	Total
1	7	29	205	29	452	13108
2	10	42	426	20	318	6360
3	12	52	659	16	255	4080
4	22	88	1998	9	149	1341
5	49	177	8995	4	74	296

2.3.2 Input data for flood simulations

LiDAR data have been used to create DTMs with 2.5, 5, 10, 20, 40 and 80 m grid cell size. As already seen, the study area is crossed by a road embankment that influences the hydraulic simulation: the smaller cell dimensions can better represent the embankment (fig. 3b), while the larger grid dimension (20, 40 and 80 m) corresponds to a rougher representation. With larger grid sizes, height averaging process connected to data interpolation causes an underestimation of embankment's top (fig. 3c). To overcome this problem, new DTMs have been created for each resolution, correcting the elevation of the cells located over the embankment, with new values corresponding to embankment actual height (fig. 3d).

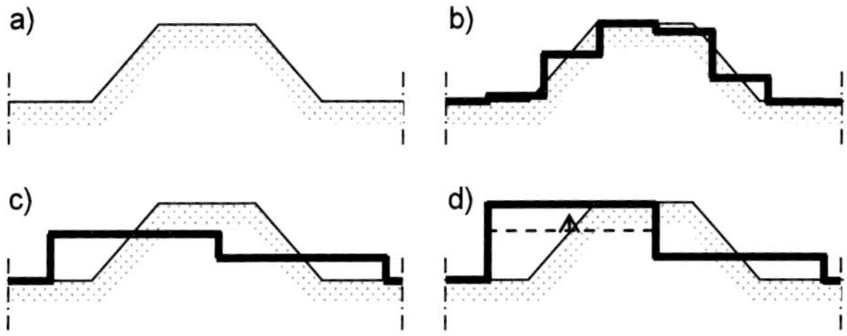

Figure 3: Cross section schematization of road embankment (a) and grid model representation: b) high resolution DTM, c) low resolution DTM and d) low resolution DTM modified.

2.4 Boundary conditions

For hydro-morphologic modelling, a flood with $T_R = 20$ years was used as inlet boundary condition (peak discharge 459 m^3/s) and a variable water depth calculated according to the hydrograph was set at the outlet section. A null value of bed level change was assumed both upstream and downstream, in order to avoid numerical instabilities. Sediment transport at the inlet was considered to be zero, due to the presence of a dam at the beginning of the reach.

Solid transport has been described exclusively as bed load, due to gravels movement, estimated with the empirical equation of Smart and Jaeggi [10]. Three sediment fractions (15, 30 and 50 mm) were used to describe grain size distribution in the channel, and two homogeneous reaches were defined with different percentages of these fractions.

For flood modelling, two series of hydraulic simulation (each one considering the six grid resolutions) have been performed: the first series uses the DTMs automatically generated from LiDAR points, while the second one uses DTMs modified to simulate the presence of the road embankment. For simulations run with modified DTMs, the 1D flow module is also applied, to represent the drainage channel.

For each simulation, the water level at the end of the drainage channel (upstream the pump station) has been used as a downstream boundary condition, while a triangular shaped inflow hydrograph ($T_R = 50$ years) has been set as an upstream boundary condition.

A constant value of Manning's roughness (0.04) has been set for all simulations according to previous analyses (Fattorelli and Frank [11]).

2.5 Results

The impact of topographic information on hydraulic and morphological model results was evaluated for both areas through a comparison of results obtained for different input data.

Regarding hydrodynamic results, the following comparisons are presented: liquid peak discharges at the outlet of the studied reach, lag time between upstream-downstream peak discharges, maximum water levels in a representative portion of the river reach, and morphologic variations and position of erosional and depositional zones along the watercourse.

For flood simulation, analyses of flooded area dynamics and extent of flood upstream and downstream the road embankment are considered.

In order to analyse the effect of model resolution on the position of erosional and depositional areas along the watercourse for the Torre river, and to evaluate the extend of the flood for the Adige floodplain, the statistical measure of fit (Horritt and Bates [4]) was used:

$$F = \frac{A_1 \cap A_2}{A_1 \cup A_2} \tag{1}$$

where $A(\cdot)$ gives the surface classified as above (deposition) and below (erosion) the threshold (± 10 cm) for the hydro-morphological model or the area predicted as inundated by the flood modelling. The indexes *obs* and *sim* indicate the reference and simulated surfaces. The results obtained with bathymetry 'a' and grid 1 and the inundation extent according to the 2.5 m modified DTM are assumed to be the reference values for the hydro-morphological model and for the hydraulic one, respectively, because they presents the best resolution condition.

The value of F indicates if the model parameters are correctly predicted, and may range between $F = 1$, in case of perfect fit (identical results are obtained) and $F = 0$, if there is no intersection between simulated and observed areas.

2.5.1 Hydro-morphological model results

Liquid peak discharges (Q_L) at the outlet of the reach, calculated with different grid dimensions and the two model bathymetries, are shown in the fig. 4a. Note that the peak discharges obtained with LiDAR topographic data, show similar values for the three first grids, and thereafter decrease. The lag time (fig. 4b) increases from grid 1 to grid 5. With decreasing grid resolution, there is a smoothing effect of riverbed irregularities, which generates larger cross-sections and lower water depths. As a consequence, slower flow and higher flood wave mitigation are produced. Considering the cross-sectional data, note that, for the different grids, the Q_L values are similar, while lag times are lower than the corresponding ones calculated with LiDAR data. These tendencies may be

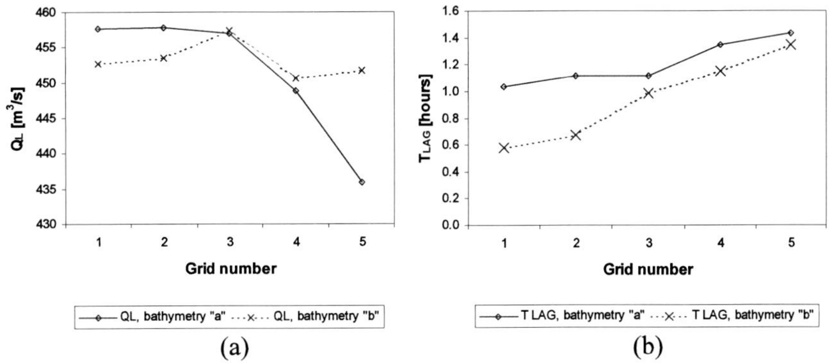

Figure 4: Comparison of: (a) liquid peak discharges downstream of reach; (b) lag time for hydrodynamic simulation with different resolution data (grid and model bathymetry).

explained by the raw bathymetry effect, due to the lack of data between cross-sections. This generates a bathymetry with fewer curves (straight channel) and does not describe accurately the morphological features (islands, bars), resulting in a lag time that, for grid 1, is almost half that calculated with bathymetry 'a'. The same effect due to grid size described above for bathymetry 'a' is also observed in this case.

An example of simulated maximum water level, with different resolutions of topographical model and grid dimensions, is shown in fig. 5. The greater dissimilarity is produced with model bathymetry 'b', originating from the traditional topographical survey. In this case, using the largest cell dimension (grid 5) causes uniformity of maximum water levels, which is greater than the smallest cell dimension (grid 1). This is clear considering that the riverbed is more homogeneous and morphological bed details like central and point bars are lost (fig. 5).

Considering the position of erosional and depositional areas along the watercourse for the Torre river, values of F (Eq. (1)) obtained with bathymetry 'a' for grids 2-3 (fig. 6) are close to 0.6, showing good spatial representation of erosional-depositional areas, while with grids 4-5, the F values decrease to 0.29-0.20, indicating that, in those cases, the results are not able to describe the spatial distribution accurately. Obviously, these results are due to the grid dimension effect, as discussed above.

The F values obtained with bathymetry 'b' (around 0.20) indicate that, for all grids, even the one with the highest resolution, the model cannot identify the spatial distribution of erosional-depositional areas with enough precision. This, of course, is due to the lack of data between two consecutive cross-sections.

The results obtained with the above statistic can also be observed in fig. 7, in which the erosional-depositional areas are similar for simulation done with grids 1 and 3 with both bathymetries, but they are highly different when using grid 5 as input. It is also important to highlight the great difference between grids 1 and 5 on the area spatial distribution obtained with both bathymetries.

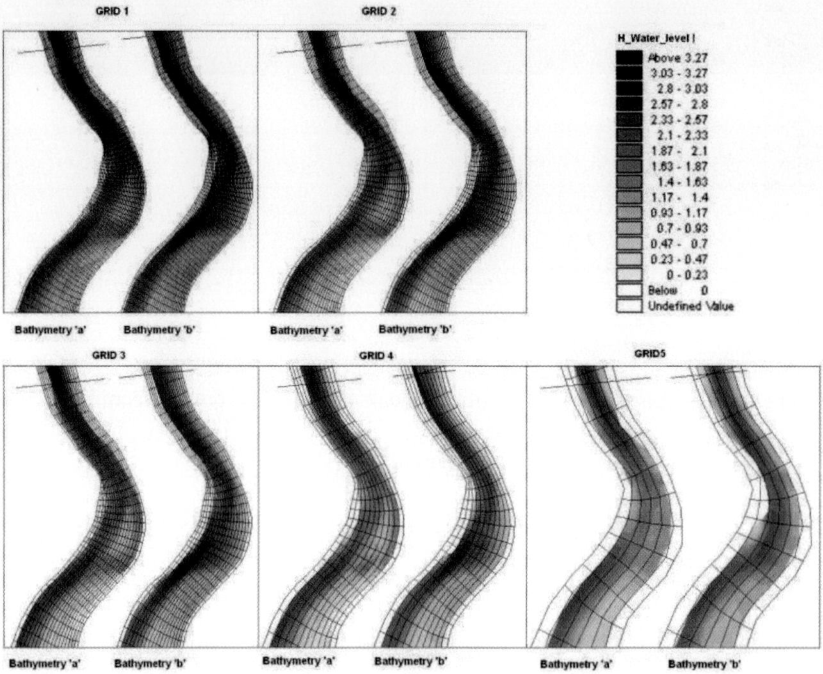

Figure 5: Comparison of simulated maximum water depth with different topographical resolutions (bathymetries 'a'-'b') and cell sizes (grids 1 to 5).

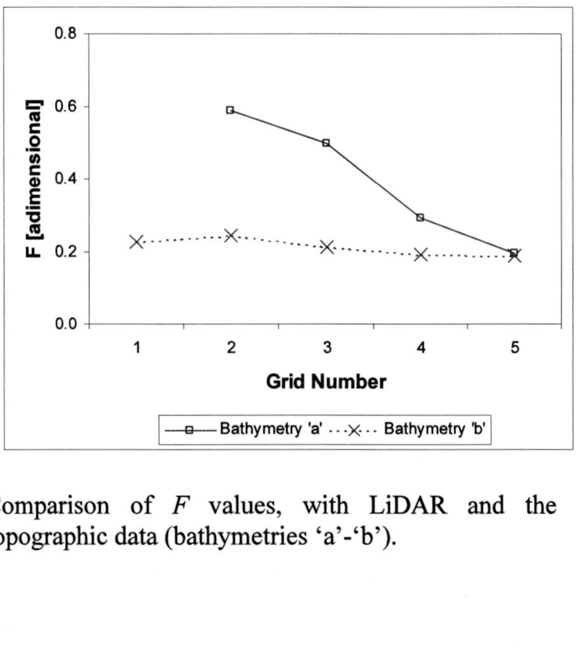

Figure 6: Comparison of *F* values, with LiDAR and the cross-section topographic data (bathymetries 'a'-'b').

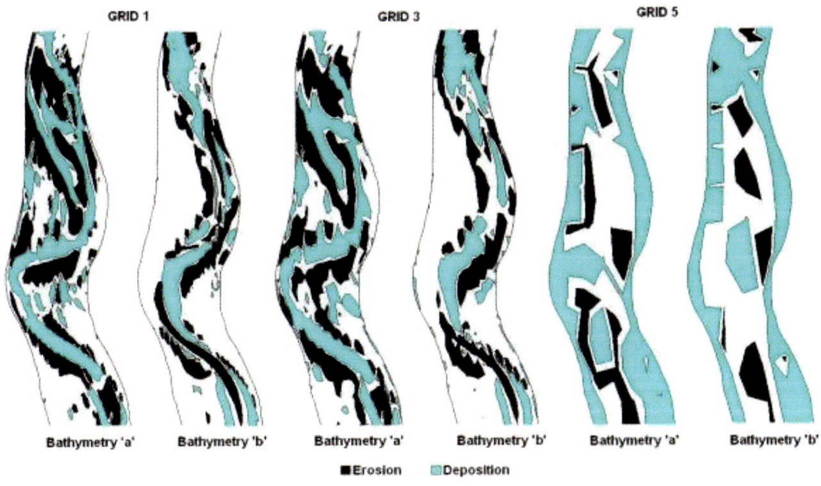

Figure 7: Erosional-depositional areas in a portion of the watercourse obtained with grids 1, 3, 5, and bathymetries 'a'-'b'.

2.5.2 Flood modelling results

In simulations run using only 2D module, drainage effect due to channel presence is correctly represented only with high resolution DTMs. With coarser DTMs (10, 20, 40 and 80 m), the water boards on the plain permanently, since the drainage channel is not detected due to cell dimensions (fig. 8). DTMs modifications and schematization of the drainage channel through the 1D module, assure better drainage dynamics' simulations for each resolution (fig. 9).

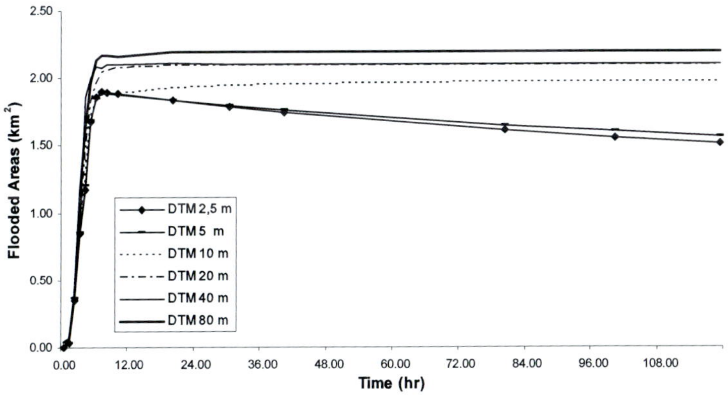

Figure 8: Inundated areas dynamics for different resolution of automatically generated DTMs.

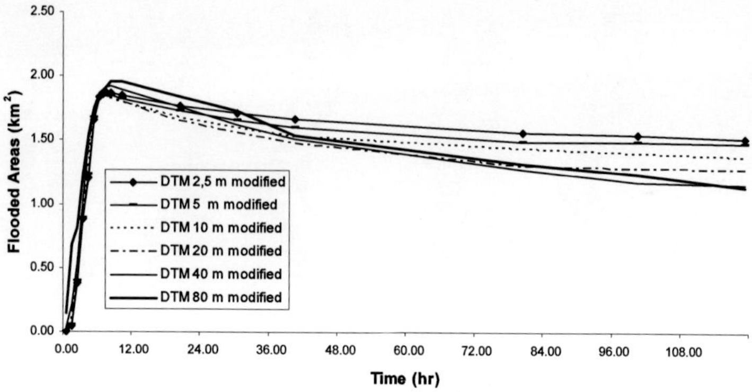

Figure 9: Inundated areas dynamics for modified DTMs.

Considering flood modelling, modification on DTMs slightly influence inundated area upstream the embankment, while downstream areas variation harder reflect road embankment presence and embankment detection (fig. 10). Outputs derived from simulations run with 20 m modified DTM are similar to outputs derived from 2.5 and 5 m automatically generated DTMs.

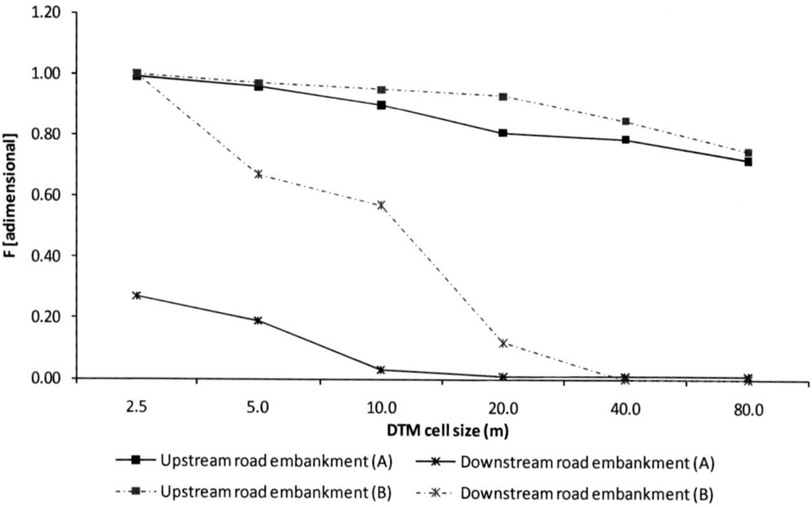

Figure 10: Comparison of F values, for different DTM grid size for automatically produced DTM (A) and modified ones (B)

2.5.3 Simulation time

The simulation time using large cells for hydro-morphological and hydraulic simulations change drastically, according to the grid size. For the MIKE21C simulations, computational time is 1.5 minutes for grid 1and increases with grid resolution up to 33 minutes for grid 5. For flood simulation, simulation takes about 15 minute for the 80 m grid, while the 2.5 m DTM required days of computational time. This may become critical when dealing with diverse hydrological scenario, corresponding to flood waves of different return time, T_R, has to be simulated, resulting in a simulation time of several days.

3 Conclusions

When a hydraulic or a hydro-morphological model is applied, the availability of topographic data influences the results that can be obtained; the dimensions of the cells should be chosen according to the morphological features present on the study areas.

The analysis carried out in this study, applying the bi-dimensional hydro-morphological model MIKE21C and the mono-bidimensional model Sobek, indicates that resolution topographic data and cell dimension models produce considerable dissimilarities in simulation results; both on morphological variations (distribution of sediment transport patterns, erosion rates) and hydraulic results (lag time, water depth, flood extent). Using models based on high-resolution LiDAR data, accurate simulations of hydraulic and hydro-morphological variables were obtained.

A comparison between model results obtained with different grid dimensions highlights important ground feature loss, mainly derived from averaging and internal interpolations of elevation data, which influences model performance and reliability of simulation results.

The analysis on the flood simulations, also, revealed that the use of detailed topographic information itself is not enough to ensure a correct hydraulic representation of surfaces, since this representation depends upon DTM resolution and it is strictly connected to computational time that needs to be consistent with study requirement.

This study indicates that optimum choice for grand-scale analysis could be a low resolution DTM processed to represent correctly natural or artificial features (road embankments, minor water streams) that strongly influence flood paths. Concerning embankments (roads, railways), effectiveness on representation of top height is essential, concerning drainage network, use of coupled 1D-2D models is the key for model reliability. In case of detailed analysis, instead, it is necessary to choose a digital terrain model with a small cell dimensions; DTMs resolution should be suitable to guarantee a greater reliability of ground feature representation.

For hydro-morphological simulations, raw bathymetric data generated from field surveys (cross-sectional data) produced a smoothing of topography between cross-sections and a consequent loss of morphological details, which does not allow accurate definition of the position of erosional-depositional areas.

Larger grid dimensions reveal as well a gradual loss of detail in the description of topography, and therefore, difficulties in reproducing the characteristics of morphological evolution. Instead, small cells accurately described sediment transport processes or feature that might influence flood dynamics, but they increase the calculation time.

Analysis of the spatial distribution of erosional-depositional areas simulated using LiDAR input data shows that up to 16 cells (grid 3) used in the cross-sectional direction, gives sufficiently accurate results. This may be explained by considering that, in a braided river, it is essential to have a minimum number of cells on the cross-sectional direction in order to describe each of the existing morphological features (bars, islands, multi-channels, etc.). A lower number of cells (grids 4 and 5) do not describe all these features well enough.

References

[1] Brath, A. & Di Baldassarre, G. *Effetti del grado di dettaglio dell'informazione topografica nella simulazione numerica bidimensionale*. L'Acqua, 1, pp. 25-30, 2006.

[2] Casa, A., Benito, G., Thorndycraft, V.R. & Rico, M. *Efectos de las fuentes cartográficas en los resultados de la modelación hidráulica de crecidas*. Ingeniería del Agua, 12(4), pp. 309-320, 2005.

[3] Haile, A.T. & Rientjes, T.H.M. *Effects of LiDAR DEM resolution in flood modelling: a model sensitivity study for the city of Tegucigalpa, Honduras*. Workshop "Laser scanning 2005" Enshede, Netherlands, pp. 168-173, 2005.

[4] Horritt, M.S. & Bates, P.D. *Effects of spatial resolution on a raster based model of flood flow*. Journal of Hydrology, 253, pp. 239-249, 2001.

[5] Lane, S.N. & Richards, K.S. *High resolution, two-dimensional spatial modelling of flow processes in a multi-thread channel*. Hydrological Processes, 12, pp. 1279-1298, 1998.

[6] Marks, K. & Bates, P. *Integration of high-resolution topographic data with floodplain flow models*. Hydrological Processes, 14, pp. 2109-2122, 2000.

[7] DHI (Danish Hydraulic Institute). *MIKE 21C, scientific documentation*. 2005.

[8] Wright, N.G., I. Villanueva, P.D. Bates, D.C. Mason, M.D. Wilson, G. Pender, and S. Neelz, *A Case Study of the Use of Remotely-sensed Data for Modelling Flood Inundation on the River Severn, UK,*. Journal of Hydraulic Engineering, 134(5). 2008.

[9] BETA Studio srl. *Progetto preliminare per il ripristino dell'officiosità idraulica del torrente Torre mediante modellazione idraulica dell'asta del torrente torre dalla diga di Crosis, in comune di Tarcento, fino alla confluenza col Fiume Isonzo al fine della messa in sicurezza del territorio.* Regione Autonoma Friuli - Venezia Giulia. Protezione Civile della Regione, OPI CD2/430.064, 2006.

[10] Smart, G.M. & Jaeggi, M.N.R. *Sediment transport on steep slopes*. Mitteilung der Versuchsanstalt für Wasserbau, Hydrologie und Glaziologie der ETH Zurich, 64, pp. 19-76, 1983.

[11] Fattorelli S., Frank E. A distributed technique for flood damage assessment using GIS and a 2D hydraulic model. River Basin Management III, WIT Press, Southampton (UK), pp. 433-442. 2004.

An investigation of the relationship between perceptions of social responsibility and community resilience to flood

A. Mullins & R. Soetanto
Department of Built Environment, Coventry University, UK

Abstract

Despite well-documented evidence of the potential physical impacts of flooding, research has so far neglected to fully investigate the manner by which different aspects of decision making at community level could influence resilience to flooding. This research attempts to investigate this gap in knowledge by exploring ways in which a better understanding of the expectations associated with social responsibility could potentially increase community resilience. There is particular emphasis upon the interrelationships between social responsibility and the decision making process. This paper sets the context for this investigation and proposes a methodology that attempts to not only understand how key community groups perceive their own levels of social responsibility to the community, but also what level of social responsibility they believe the other groups should have. It is argued that community groups may not even be aware that they are failing to meet their expected level of social responsibility. Therefore, the gaps discovered by this methodology between a group's own perception and how they are perceived by others would highlight barriers to community resilience. An argument for research to better understand resilience at the level of the community by exploring the individual and interconnected decision making of householders, small businesses and policy makers, is further elaborated. The arguments presented here will be of interest to community leaders and provide considerations for built environment professionals embarking on the development of resilience measures, with considerations suggested for future research within this field.
Keywords: climate change, community resilience, decision making, risk perception, social responsibility.

1 Introduction

The academic literature indicates that human activity is having a large, detrimental effect upon the environment, increasing climate change and thereby increasing the likelihood of extreme weather events, such as severe flooding [1]. In modern times, our built environments have become increasingly merged with the natural environment, making both more susceptible to flooding. Flooding has not only become more frequent and more severe, but also society has become more vulnerable to its effects. Flooding is of particular concern within the UK, and in 2007 there was widespread flooding that caused an enormous amount of damage as again our fragile infrastructure was not able to cope with such an extreme event. There is an inextricable link between people, their built environment and its relationship with flooding. The better prepared people within a built environment are, the smaller the impact the flooding is able to have. This is important because above all the forecasts and technologies of the modern age, it is argued that the people remain the key to a successful defence. Following the extreme flooding event in 2007, the Pitt review [2] acknowledged the importance of building resilience at local community level, and provided tangible evidences of how this resilience could limit the damages from flooding. However, research has so far neglected to fully investigate this area within the built environment with which we are most familiar and is most salient to our needs: our own community. This paper reports an investigation to enhance understanding of decision making processes and interrelationships between three key community groups (policy makers, householders and small businesses) in order to improve the resilience of the local community. A better understanding of social responsibility, the decision making process, and interrelationships amongst members of the community will help joined-up thinking and optimise the selection of adaptation and mitigation strategies to flooding. Firstly this paper discusses resilience and decision making at the level of the community, before exploring potential barriers and drivers for community resilience through a greater understanding of social responsibility. This paper details the proposed methodology for investigating this area and concludes with considerations for further research.

2 Community resilience

Many definitions of resilience describe communities dealing with the effects of a flooding event and then returning to their normal functioning prior to the event. However, if a community returns to its previous state then it may have bounced back from the event but it has not actually increased its resilience to similar events. Instead, resilience must be thought of as containing elements of learning and adaptation to events so that community resilience can be increased. The resilience of a community is determined by the interconnected system's ability to absorb disturbance, self-organise and contain the capacity to learn and adapt [3]. It is the attitudes, perceptions and behaviours that members of a community adopt or display prior to a flooding event that can determine the ability of that community to absorb the disturbance. Furthermore, these aspects also determine their motivation and ability for self-organisation during the event and how much they are willing to learn from the event in order to change their perceptions and behaviours.

There has been research conducted on a number of aspects of extreme flooding events and climate change, such as resilience, adaptive capacity and vulnerability at the national, regional and sector levels [4], but there has been very little research conducted within the heart of our built environments, at the local community level. As a conceptual framework, it is helpful to understand a community in geographical terms as the members of these communities not only share the resources of that area but also have a shared risk of hazards. Furthermore if members of these communities share common resources and hazards it may be easier to identify the differences between individuals that possess different levels of engagement with the issue of climate change. A localised approach will provide a better context for understanding the decisions of members of the community who fail to engage in resilience promoting actions. Lorenzoni and Pidgeon [5] support this view, stating that although there is concern regarding climate change present in Europe and the USA; it is not a high enough concern to change behaviours in daily lives and therefore saliency of risk must be increased by concentrating communication of risk at the community level.

There are four main stages to the resilience process, collectively known as the social resilience cycle [6]. The first stage is mitigation where there is a general process of increasing a community's ability to cope with a flooding event, for example by not building on flood plains or by better protecting buildings. The decisions associated with this stage are the planning and preparation decisions made before the flooding occurs, such as training staff, which provide a basis for community resilience to the flooding event. This first stage is arguably the most crucial stage in determining the degree of resilience that a community will have to a flood as it can also affect the capabilities of the later stages. The first stage is also the phase where perceptions, beliefs and other human barriers can create the most diverse behaviour, as trying to convey the dangers of a flood which has not yet occurred is infinitely more difficult than pointing out the danger and destruction that surrounds people in the later stages. Therefore, these potential barriers to resilience need to be better understood.

2.1 Key community groups

The continued successful resilience of the community in the short to medium term relies upon the groups which make up that community, such as the householders, small and medium-sized enterprises (SMEs) and policy makers. Communities are made up of individuals, each of whom can have an effect upon their personal level of resilience to flooding, which in turn will have an effect upon their community resilience. Individuals have a responsibility then to increase their resilience and they can do so through their lifestyle choices and the decisions they make about being aware of the risks faced by their community. Unfortunately, many people are unaware or are in denial about the risks they live with each day. It is these counterproductive attitudes and flawed decision making which needs to be changed in order to increase resilience. In order to instigate the necessary changes, researchers need to firstly understand how and why people reach the decisions they do about the risk of flooding, as well as understanding how the interdependencies within the community and societal infrastructure as a whole can affect these decisions. For example, why do local policy makers make the decision to build houses on flood plains when they know that this decreases their community resilience to an extreme flooding event? Why do householders and businesses make the decision to occupy buildings on flood plains when they know that this decreases their personal resilience to an extreme flooding event?

The example above indicates that there is a lack of understanding of individual and social responsibility being taken for actions that can affect personal, community and national resilience to flooding, and tension between competing factors influencing decisions. We may live in a modern blame culture but there appears to be a lack of accountability for the tragedies that occur when the effects of disasters are increased because individuals have to make less than optimum decisions that have decreased their resilience to such events. Is it the fault of householders who choose to live there or the fault of policy makers who allow people to build there? Too often flooding events are blamed on being an 'act of God' when in fact a clear pathway of poor decisions made over a long period of time have contributed to the final damage caused by these events. Furthermore the over reliance upon others that is fostered through our modern interdependent lifestyles can also contribute to attitudes, decisions, expectations and behaviours which are detrimental to our resilience. The over reliance upon the emergency services can leave vulnerable situations throughout the community. Therefore, individuals need to play a greater role in increasing both their personal and community resilience to ensure that in the future communities will be better protected against these events. However, before we can change behaviour we must first overcome the perceptual barriers that affect our decision making.

3 Perceptions of risk in decision making

In the US, personal responsibility is recognised by the Federal Emergency Management Agency as being the key to building a resilient community. It has been shown though that households, SME's and policy makers underestimate risks that appear distant or global, such as the risk of extreme flooding events which are rare [7]. Communities not fully acknowledging the information available and thereby not acknowledging the risk or understanding their roles and responsibilities has been seen in studies in the USA, Canada and Europe [8]. Particular community groups may not even acknowledge that they have any roles or responsibilities towards flooding events or community resilience at all, as even simple denial of the risk has been found to justify lack of action on climate change [9]. There are many different perceptions of the risks posed by flooding, with some people suggesting that immediate action should be taken, others suggesting that the scientific evidence is unreliable, or given the uncertainty nothing should be done until there is more reliable evidence, or simply not believing that climate change affects their lives in any way [5]. Therefore, how much people believe they are at risk of flooding may determine their behaviour in preparing against such a disaster. Given then that perceptions of risk are not well understood or even accepted by many community stakeholders then it is reasonable to suggest that perceptions of roles and responsibilities relating to this risk may contain both conceptual and behavioural aspects which are detrimental to community resilience.

This makes the issue of flooding a particularly complicated concern for people making decisions as individuals, organisations, communities and society as a whole. Given that modern society contains masses of interdependencies to function efficiently, it will require an appreciation of further interconnectedness, in the form of joined-up-thinking between key community groups, to efficiently increase community resilience. This need for integration is reflected in community resilience models [10] which have stressed the importance of participation and the need to integrate stakeholders. This is further emphasised by the need to integrate community groups within climate change education. Top down information (policy makers telling people what should be done) does not appear to be working and therefore bottom up information (community groups integrating information together) is suggested to improve risk communication and community resilience [11]. Several main activities of Local Resilience Forums in the UK regions are to raise awareness of flooding risks, to enhance flood warnings before flooding, and to provide information of welfare needs once flooding has receded, at local community level [2]. However, there are many emotional and psychological obstacles which hamper interconnected decision making and resilience measures. These support the overall view that it is the decisions we make about our behaviour, whether it be as individuals or as part of a team, which creates the largest obstacle to successful measures to deal with flooding. The decision making process of individuals is recognised as being a vital part of community resilience. However, a number of factors can negatively affect the decision to engage with the issue. Therefore it is imperative that research identifies the most important of these factors and attempts to combat their negative effects upon the decision making process so that improved non-technical resilience measures can be proposed. We will now explore social responsibility as one of the main factors that contains the potential to affect the decision making process.

4 Social responsibility

One of the main areas to emerge from the discussion of resilience is the idea of individuals being more socially responsible and accepting a greater level of individual responsibility for community resilience. Therefore exploring perceptions of social responsibility for flooding will provide an excellent platform from which to investigate barriers and drivers to community resilience.

It is important though to understand the concept of social responsibility and how it may be used as a research tool.

4.1 Social responsibility as a concept

Social responsibility is a term that has been utilised in a variety of forms but is widely recognised as relating to the relationships between the economic, environmental and social aspects of an organisation or groups activities that endeavour to benefit society [12]. It is largely agreed that social responsibility is an important topic not only for the business environment but also for wider society, with negative effects, such as new legislation and adverse publicity, seen as arising from a failure to recognise and maintain a suitable level of social responsibility [13]. Understanding how people perceive themselves and each other in relation to a particular aspect can be a useful way of investigating that aspect itself. This raises the question of how do these three key community groups view each other's roles and responsibilities and are there any gaps between expectations of others and understanding of one's own role? These gaps would represent potential barriers to increasing community resilience.

4.2 Social responsibility as a research tool

When investigating social responsibility research has largely focused on how businesses attend to societal needs through corporate social responsibility (CSR), however it could be argued that this has largely been an investigation of public relations rather than actually exploring the processes associated with social responsibility. Corporate social responsibility and public relations share such strong similarities in their origins, theories and practices that the distinction between the two fields has become blurred. It has even been stated that public relations is simply the practice of social responsibility, despite there being key differences between these two fields [14]. Therefore when one thinks of social responsibility they think of the responsibility that businesses have to the general public and how they communicate information to the public and act upon the feedback [15, 16], however this may actually be a more fitting description of the foundations of public relations models, such as the four step management process [17] and the RACE framework [18], rather than social responsibility. Even the foundations of CSR models themselves, such as the four-step process of corporate social involvement [19], may not be suitable to investigate the relationship between social responsibility and community resilience. This is because CSR models are built with the purpose of being related to the business, with the public being a part of this particular business process. CSR is influenced by a number of driving actors, such as investors, consumer demand, government regulation, supply chain requirements and civil groups, all of which apply in varying degrees to different businesses [14].

However, when investigating community resilience it is not solely the community group's responsibilities to each other which is being investigated, but is instead largely focused upon their responsibilities to the community itself and their roles within it. This is an important distinction that highlights the differences between social responsibility as an independent aspect, rather than CSR, which is a business orientated view of social responsibility, and public relations models, which although do allow a two-way flow of information are not suitable for climate change research as they do not provide true equality and integration between multiple community groups as again they have been created for a different purpose. It is unknown therefore whether or not the drivers identified for social responsibility in a corporate context will apply to perceptions of social responsibility in relation to climate change and this paper proposes a different use of social responsibility as a research tool.

Given then that community resilience to flooding relies upon the successful integration of each of the three key stakeholder groups, householders, SMEs and policy makers, then it is reasonable to suggest that social responsibility research should not be conceptualised or investigated as a circular process (see figure 1, illustration on the left), as this limits integration, which this paper suggests is a necessary component of future resilience measures. Social responsibility research instead needs to investigate perceptions of the roles and responsibilities that the key community groups have not only of themselves, but also how they perceive the other groups, with new ideas generated and communicated by each of the groups rather than the public simply providing feedback on business ideas or policies, creating a multi-path framework of perceptions and providing a basis for integration (see figure 1, illustration on the right).

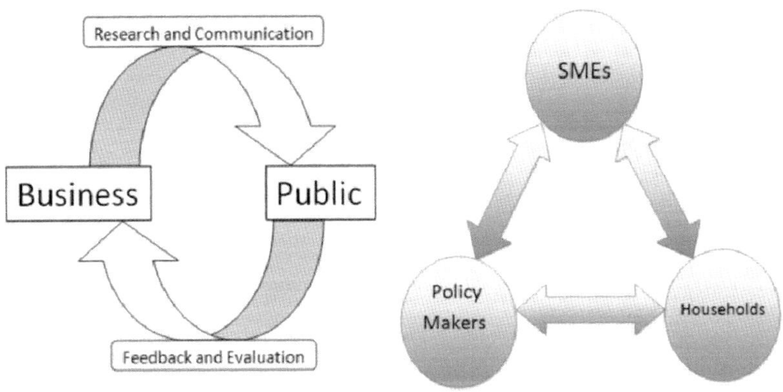

Figure 1: Public relations process vs. community social responsibility framework.

Exploring social responsibility in this integrated manner will highlight potential links between these community groups, how they are contextualised by social responsibility and how this may affect overall community resilience. For example, it is reasonable to state that householders may expect policy makers to do everything they can to prevent flooding and policy makers may expect householders to do everything they can to lessen the impact if it does flood. This point was also raised by the Pitt Review [2] in that the responsibility does not rest solely on the UK Government and other relevant authorities as they cannot protect the community from all consequences of disasters. However history has shown us that householders do not do anything until it is too late, such as ignoring flood warnings due to experience of false alarms, and when it does go wrong they then shift the responsibility to the policy makers. But the policy makers have to follow procedures which often assume that the householders are taking actions to lessen the impact of flooding. It is these kinds of gaps and misunderstanding of social responsibilities that can cause failings in resilience measures and drain extra resources. The householders are blaming the policy makers when in fact they have decreased their own resilience (by not taking actions to protect themselves) and their community's resilience (by allowing the flood to cause greater damage and thereby using up more of the limited resources available). Current perceptions of social responsibility need to be understood to identify these gaps and misunderstandings so that attitudes and expectations can be changed and resilience improved. These gaps would not be highlighted simply by only asking each group about their perceptions of their own responsibilities and therefore this integrated approach needs to be adopted.

5 Methodology

The methodology proposed is designed to explore perceptions of social responsibility within and between community groups, with the objective of highlighting potential barriers or drivers for community resilience. The investigation will use a mixed methodological approach, with questionnaires providing quantitative data regarding perceptions of social responsibility and open-ended interviews providing qualitative data. This design allows the questionnaire responses to provide an overview of perceptions of social responsibility within and between community groups and the interview responses to provide more specific details about the relationships between community groups and place the broader perceptions in context.

The study will use participant information sheets and consent forms for the questionnaires and interviews to provide details about why the research was being conducted, what was expected of the participants, providing contact details of the researchers and informing participants of their rights regarding participation and data use. The study will use four versions of a Perceptions of Social Responsibility Questionnaire, one containing questions about the self, one relating to householders, one to SMEs and one to policy makers. These questionnaires are based upon a modified version of Berkowitz and Lutterman's [20] Social Responsibility Questionnaire which has provided a valid and reliable basis for researching social responsibility since its creation. Modified versions of the original questionnaire have been used in research informing social responsibility scales [21], exploring ethics and social responsibility in relation to grocery shopping [22], testing attitudes in relation to social involvement models [23] and exploring psychosocial factors that influence volunteer work [24].

In the same way that Berkowitz and Lutterman [20] describe participation of individuals in society as leading to greater adoption of that society's attitudes and values, so too can it be reasonably expected that a householders, SME manager's or policy maker's role within a community be likely to lead to adoption of community norms. This is achieved through both laws and social rewards for meeting the expectations of those roles within the community; however it is unknown exactly what perceptions and behaviours these expectations create within the mindset of each community group for any given aspect of the community, for example during an extreme flood within the community. The original Social Responsibility Questionnaire measured an individual's acceptance of the traditional values of their society. The aims of this project though are to reflect the perceptions of a community group in relation to a particular aspect and as such the original attitudinal statements were modified to meet the aims of the research.

All four versions of the modified Social Responsibility Questionnaires will use the 12 modified attitudinal statements, with the terminology regarding the 'self' in the self-perception questionnaire being directed towards householders, SMEs or policy makers in their respective questionnaires. The statements will be scored using a 4 point Likert scale ranging from Strongly Agree (4) to Strongly Disagree (1), with a number of statement being reverse scored to counter acquiescence. This will give each questionnaire a potential score range of between 12 (representing very low social responsibility) and 48 (representing very high social responsibility). These attitudinal questions will provide information about how each community group views their own social responsibility and how socially responsible they perceive the other two community groups to be.

In order to provide a context for the perceptions of social responsibility highlighted by the questionnaires, an open-ended social responsibility interview schedule will be designed to facilitate cognitive mapping interviews. Cognitive mapping is a technique used to structure and evaluate accounts of problems and previous research has utilised cognitive mapping to explore decision making processes at both a micro level for individual problem solving [25] and at a macro level for corporate strategy development [26], as well as to investigate related issues such as risk [27]. Cognitive maps are a widely used validated research tool for exploring representations of knowledge of particular subjects, problem solving, decision making and representing attitudes [28].

Cognitive mapping will be used as a tool to record information gathered from interviews, achieved by recording phrases used by the interviewees under particular headings. These headings become concepts which are presented in a visual format displaying their relevant connections and interactions, revealing patterns of reasoning [26]. Cognitive mapping produces a representation of how an individual views a particular problem topic, in this instance their own or others social responsibility. It also involves the noting of opposite poles of information to help explain the meaning of particular concepts and the identification of possible options and outcomes within pairs of concepts, highlighting conflicts between different individuals [26]. Furthermore, the grouping of cognitive maps also allows individuals to see where their view stands in relation to others, increasing deeper understanding of the topic and highlighting gaps or potential alternatives to existing measures [26]. Cognitive maps of the resilience of individual community groups could be merged to create a collective map of community resilience. These qualities make it a useful tool for exploring perceptions of social responsibility both within individual community groups and between community groups.

6 Conclusion

Climate change is a serious threat and extreme weather events, such as severe flooding, are becoming more frequent within the UK. It is important that communities find new ways to increase community resilience to such events and it is proposed that the three community groups that will play a key role in this aspect are householders, SMEs and policy makers. Current community resilience models and research in the area have highlighted social responsibility as a potential factor that affects resilience levels and this research proposes that further exploration of this aspect may highlight potential barriers and drivers for community resilience. This will be achieved by using an inter-relational framework for exploring social responsibility within and between these three key community groups. With this, we can investigate current self-perceptions of social responsibility of key community groups, investigate current perceptions of social responsibility between key community groups and explore the relationship between perceptions of social responsibility and community resilience in relation to flooding. This paper proposed a methodology which investigates social responsibility in this way. The gaps between the expectations a community group has of meeting its own social responsibilities and the expectations other groups have of that group will be indicative of barriers to resilience. This is because either social responsibilities are not being met by that group or are being wrongly attributed to that group within the perceptions held by the other groups. The results will provide a deeper insight into the affect that different perceptions of social responsibility may have upon the wider issue of community resilience. It is believed that the findings will further highlight the need for integration between community groups and will be of use to community leaders (e.g. Local Authorities and Local Resilience Forums including various local flood action groups) to make better decisions regarding community resilience measures.

Future research in this area must consider that many social and psychological factors may not be distinct from each other and may influence and affect each other, as well as the overall decision making process, rather than simply being a correlation between the two aspects of social responsibility and community resilience. This paper has focused on the first stage of the social resilience cycle, but future research should also consider the effects that improved social responsibility may have upon the other stages. One of the first tasks faced by researchers though is establishing a common framework for measuring social responsibility within the community. There is much work to be done in this field, but what can be concluded is that the role of perceptions of social responsibility are extremely important when trying to protect our built environments from flooding disasters.

References

[1] IPCC – Intergovernmental Panel on Climate Change, Climate change 2001. Impacts, Adaptation, and Vulnerability. Cambridge University Press: Cambridge, 2001.

[2] Pitt, M., *The Pitt Review – Learning Lessons from the 2007 Floods*. London: Cabinet Office, 2008.

[3] Walker, B. & Salt, D., *Resilience Thinking*. Island Press, 2006.

[4] Gallopin, G. C., Linkages between vulnerability, resilience and adaptive capacity. *Global Environmental Change,* **16**, pp. 293-303, 2006.

[5] Lorenzoni, I. & Pidgeon, N., Public views on climate change: European and USA perspectives. *Climatic Change,* **77(1-2)**, pp. 73-95, 2006.

[6] Maguire, B. & Hagan, P., Disasters and communities: understanding social resilience. *The Australian Journal of Emergency Management,* **22(2)**, pp. 16-20, 2007.

[7] Viscusi, W. & Zeckhauser, R., National survey evidence on disasters and relief: Risk beliefs, self-interest, and compassion. *Journal of Risk and Uncertainty*, Springer, **33(1)**, pp. 13-36, 2006.

[8] Bord, R. J., O'Connor, R. & Fisher, A., In what sense does the public need to understand global climate change? *Public Understanding of Science*, **9**, pp.205-218, 2000.

[9] Stoll-Kleemann, S., O'Riordan, T. & Jaeger, C. C., The psychology of denial concerning climate mitigation measures: evidence from Swiss focus groups. *Journal of Global Environmental Change,* **11**, pp. 107-117, 2001.

[10] Paton, D., Measuring and monitoring resilience in Auckland. *GNS Science Report 2007/18*, 2007.

[11] Dufty, N., *A New Approach to Flood Education*. Molino Stewart Pty Ltd: Parramatta, 2008.

[12] ISO, *ISO to go ahead with guidelines for social responsibility, Press Release 924, 29 June 2004*. Geneva: International Standard Organization, 2004.

[13] Peterson, R. T. & Jun, M., Perceptions on Social Responsibility: The Entrepreneurial Vision. *Business Society OnlineFirst*, 2007. http://bas.sagepub.com/cgi/rapidpdf/0007650307305758v1

[14] Clark, C. E., Differences between public relations and corporate social responsibility: an analysis. *Public Relations Review*, **26(3)**, pp. 363-380, 2000.

[15] Joyner, B. E., & Raiborn, C. A., Management caveats for measuring and assessing public responsibility performance. *Business Horizons*, **48(6)**, pp. 525-533, 2005.

[16] Trainer, T., Social responsibility: The most important and neglected problem of all? *International Journal of Social Economics*, **32(8)**, pp. 682-703, 2005.

[17] Cutlip, S. M. & Center, A. H., *Effective Public Relations* (5th ed.). Englewood Cliffs, NJ: Prentice-Hall, 1978.

[18] Marston, J. E., *Modern Public Relations.* New York: McGraw-Hill, 1979.

[19] Preston, L. E. & Post, J. E., *Private Management and Public Policy: The Principle of Public Responsibility.* Englewood Cliffs, NJ: Prentice-Hall, 1975.

[20] Berkowitz, L. & Lutterman, K. G., The traditional socially responsible personality, *Public Opinion Quarterly,* **32**, pp. 169–185, 1968.

[21] Reed, V. A., Jernstedt, G. C., Hawley, J. K., Reber, E. S., & DuBois, C., Effects of a small-scale, very short-term service learning experience on college students. *Journal of Adolescence,* **28**, pp. 359-368, 2005.

[22] Megicks, P., Memery, J. & Williams, J., Ethics and Social Responsibility Issues in Grocery Shopping: An A Preliminary Typology. *Qualitative Marketing Research*, **8(4)**, pp. 399-412, 2005.

[23] Frieden, J. B. & Downs, P. E., Testing the Social Involvement Model in an Energy Conservation Context. *Journal of the Academy of Marketing Science*, **14(3)**, pp. 13-20, 1986.

[24] Chacon, F., Menard, M., Sanz, M. & Vecina, M. L., Psychosocial Factors That Influence Volunteer Work: A Pilot Study. *Psychology in Spain*, **2**, pp. 108-115, 1998.

[25] Eden C., *Working on problems using Cognitive Mapping.* In S. Littlechild, and M. Shuttler (Eds). Operational Research in Management. Prentice Hall: London, 1991.

[26] Eden, C. & Ackermann, F., *Making Strategy: The Journey of Strategic Management*. London: Sage, 1998.

[27] Harris, C., Daniels, K. & Briner, R. B., Using Cognitive Mapping for Psychosocial Risk Assessment. *Risk Management*, **4(3)**, pp. 7-21, 2002.

[28] González, F. M., Morón, C. & Novak, J. D., *Errores conceptuales. diagnosis, tratamiento y reflexiones.* Madrid: Morata, 2001.

Decision-making methods for operational flood management

K. A. Wojciechowska
HKV Consultants, The Netherlands

Abstract

In 1995, because of danger of flooding, a massive evacuation in the province of Gelderland in The Netherlands took place. The process of the evacuation went well, however, the life-threatening flood did not occur. The decision was based on deterministic information, experience and expertise. A robust technique, where uncertainties (e.g. in water level forecast) are explicitly incorporated, was not used (and is still not used). There are several methods in the decision theory, which allow the inclusion of uncertainty in decision-making in an explicit way. This paper gives a description of such methods in the context of operational flood management. We focus on decision trees, decision influence diagrams and Markov Decision Processes. In the context of operational flood management, the methods usually require specification of conditional flooding probability, i.e. a flooding probability given some (uncertain) information. We present application of such methods to operational decisions like evacuation or activation of an emergency storage area. We compare the methods with respect to ease of understanding and we briefly discuss whether there is a chance to apply the methods in real life situations.

Keywords: decision-making, decision trees, decision influence diagrams, Markov Decision Processes, uncertainty, conditional flooding probability.

1 Introduction

The Netherlands is a country located in the delta of three European great rivers: the Rhine, the Meuse and the Scheldt. Over 60% of the country is situated below mean sea level or is vulnerable to flooding by sea or by rivers. Natural barriers like dunes and high grounds, and man-made constructions like dikes and storm surge barriers protect the area from flooding.

At the beginning of 1995, heavy precipitation in North France and the Ardennes led to rising of water levels on the rivers Meuse and Rhine. Consequently, two Dutch provinces, Limburg and Gelderland, were affected by the event. The situation in the province of Gelderland was especially dangerous, (Bezuyen *et al.* [1]). The condition of some dikes in the region was unsatisfactory. A breach in the dike system would cause flooding within a few hours in some valuable parts of the region. This fact and the resulting water depth could lead to life-threatening situations. The regional management team decided to evacuate part of the area. First, the evacuation was advised, shortly afterwards a call for compulsory evacuation was given (HWR [2]). The evacuation was massive, as almost 250,000 people were evacuated.

The process of evacuation went well, however, the life-threatening flooding did not occur. The evacuation decision was based mainly on deterministic information, experience and expertise (the reasons for the evacuation in 1995 were the water level forecasts, the duration of the high water levels and an assessment of the polder district that safety could not be assured due to the bad state of the dikes (Frieser [3])). Uncertainties in the information used (e.g. in water level forecasts) were not explicitly included in the decision-making process – a robust technique, where the uncertainties are explicitly incorporated, was not applied (and it is till now not in use).

The importance of uncertainties in water level forecast, during decision-making, can be explained on the basis of the Red River Flood of 1997. The flood was the result of high winter snow, river ice and unfavourable spring thaw. The event affected i.a. Grand Forks, a city in North Dakota (US). The water flowed far inland, leading to evacuation of the citizens (over 50,000 people were evacuated). In the face of potential flooding, the authorities did not decide to raise additional water defences because the maximal water level forecast did not go beyond prescribed safety levels (Verkade [4]). "The city engineer, relying on the 49-foot (14.5 m) estimate by the National Weather Service, stated that, had they been told of the potential for a 54-foot (16.5 m) crest, they could have planned for and built a number of secondary dikes that would have saved much of the city from serious damage. If someone had told us that these estimates were not an exact science, or that other countries predict potential river crest heights in probabilities for various levels, we may have been better prepared." (Glassheim [5].)

There are several methods in the decision theory, which can be applied to include uncertainty in decision-making in an explicit way. This paper gives a description of such methods in the context of operational flood management. We focus on the decision trees, decision influence diagrams and Markov Decision Processes. In the context of operational flood management the methods usually require specification of conditional flooding probability defined as a flooding probability given some (uncertain) information. The importance of such probabilities has recently increased in the Netherlands. We present an application of such methods to operational decisions like evacuation or activation of an emergency storage area. We compare the methods with respect to ease of understanding (of importance for decision makers, who do not necessarily have a mathematical background) and we discuss whether there is a chance to apply the methods in real life situations.

2 Emergency flood measures

In this part two emergency flood measures will be considered. These measures are evacuation and activation of an emergency storage area. In the subsequent part we present the application of decision trees, decision influence diagrams and Markov Decision Processes; decisions about evacuation and the activation of an emergency storage area are modelled with help of the methods.

2.1 Evacuation

In this research, evacuation is defined as a movement of people away from a danger or occurrence of the danger, i.e. flood.

In general, two time aspects are involved with the evacuation decision, namely required and available time for evacuation (Barendregt *et al* [6]). The required time for evacuation consists of the time needed for decision-making (evacuation yes/no), preparation (initiation, warning and response) and transportation out of the area. The time available for evacuation is the period between the decision-making and the moment the calamity actually begins. Theoretically, if the time available for evacuation is larger than the time required for evacuation, a complete evacuation of inhabitants of the endangered area is possible. On the other hand, if the time available for evacuation is smaller than the time required for evacuation, a complete evacuation is not achievable prior to the occurrence of the flooding.

The primary goal of evacuation is to diminish (potential) flood consequences, so as to reduce the number of injured and fatalities. However, every evacuation entails costs. Initial evacuation costs, the value of direct and indirect economic damage contribute towards the costs, Frieser [3]. Initial evacuation costs contain e.g. compensations paid out to the evacuated people (in 1995, every evacuated household received about €225 from the Dutch government (Frieser [3]). Direct economic damage arises as a result of the suspension of production in the risk area and depends on the sort of economic branch (e.g. minerals extraction, banks/insurance companies, building industry) and duration of the suspension. Indirect economic damage consists of losses to supply companies and customers, and losses caused by problems in traffic in the regions outside the risk area.

2.2 Emergency storage area

An emergency storage area is a sparsely inhabited within-dikes area, which can be flooded in a controlled way, if extremely high river discharges occur (RBOS [7]). The aim of an emergency storage area is to decrease the peak of a high water wave and to consequently prevent flooding in within-dikes areas, which are situated downstream from the emergency storage area and have large socio-economical importance. In short, an emergency storage area is an area flooded in favour of other area(s). The costs of activation of an emergency storage area resolved to damage to the area caused by the inflowing water.

Usability of an emergency storage area is restricted by volume of the area, meaning that the dike failure mechanism overflow is primary influenced by the controlled flooding. Influence of activation of an emergency storage area on other dike failure mechanisms (e.g. piping) is insignificant (RBOS [7]). In general, a human indirectly and directly determines the moment of activation of an emergency storage area. Indirectly, when an emergency storage area begins to work as soon as a fixed water level is exceeded, and directly, when an emergency storage area is activated by using a moveable inlet or by digging a breach. It is difficult to determine the best moment of activation of an emergency storage area, since such moment depends on partly an uncertain form of the discharge wave. It can happen that an emergency storage area is activated too early or too late. In Figure 1, two very theoretical examples are given: in the first and second panel correct and incorrect activation of an emergency storage area are depicted respectively (the modelling is done according to Nortier and de Koning [8]).

The continuous line in Fig. 1 represents a theoretical discharge wave at the Rhine (location Lobith; the river enters the Netherlands close to Lobith). The peak of the wave is equal to 17250 m^3/s; on average, this value is exceeded once per 3200 years at Lobith. The dashed line represents the discharge wave with a working emergency storage area. The area, with a volume of 161 mln m^3, is

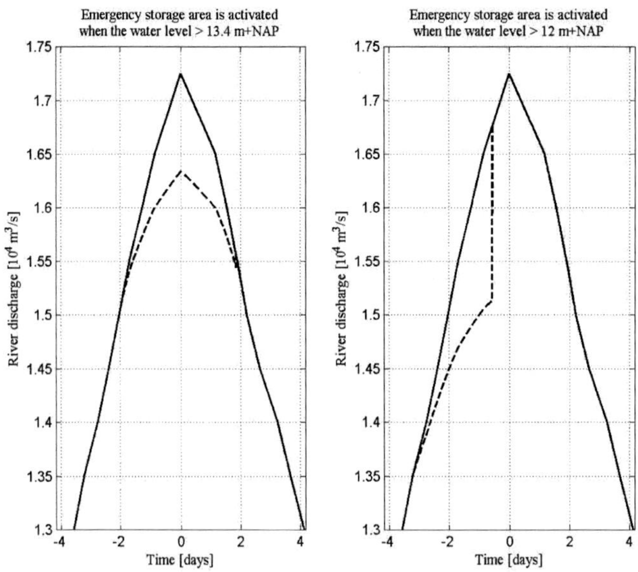

Figure 1: Influence of activation of an emergency storage area on the discharge wave.

situated in the region. The measure is activated using a moveable inlet; it is elementarily assumed that the storage capacity of the area cannot be exceeded.

In the first panel of Figure 1, the area is activated when the water level at the location is higher than 13.4 m+NAP, leading to clear reduction of the peak. In the second panel, the area is activated when the water level exceeds 12 m+NAP. Such operation does not lead to reduction of the peak; in this situation the storage capacity of the area is used up before occurrence of the peak. In other words, the main danger (so the peak) was not diminished by the activation of the emergency storage area.

3 Decision models

In this part three mathematical decision models, often invoked in the decision theory, will be considered. These models are decision trees, decision influence diagrams and Markov Decision Processes. Both theory and application of the models will be presented.

3.1 Decision trees

3.1.1 Definition

Decision tree is a method, which helps in a decision-making involving high costs and probabilities of unwelcome events. The decision tree method uses a visual approach to compare considered decisions and to assign values to those decisions by combining probabilities and costs into specific values. In particular, these specific values are expected values.

A decision tree consists of branches and nodes. In general, the first node branches out into several branches, which represent decisions. Once a decision has been made one of several states of nature takes place; each state occurs with some probability. Furthermore, costs are associated with each decisions-state pair. The optimal decision is chosen on the basis of minimum expected costs criterion. More precisely, expected costs are determined for every decision and the decision with the lowest expected costs is indicated as the optimal decision. It is worth mentioning that in the case of a complicated decision problem, the corresponding decision tree usually has a very extensive form. For more theoretical information on the subject see Benjamin and Cornell [9].

3.1.2 Application

Let us consider application of the decision tree tool. Suppose that unfavourable weather conditions are forecasted. The forecasts are, however, uncertain and the uncertainty increases with time. The weather conditions can lead to flooding in a region. A decision maker, concerned about the weather forecast and condition of dikes in the region, considers at the present time evacuation of the region or activation of an emergency storage area (located upstream from the region). On the other hand, the decision maker knows that more exact information about the weather conditions will be available in the near future, hence eventual postponing of the decision is taken into account. Furthermore the decision maker is aware that the time aspect is fundamental in the case of the considered two flood measures.

A decision tree, proposed to model the situation, is shown in Figure 2 (a version of the tree is also given in Frieser [3]). The decision tree presents possible scenarios. One of the scenarios is the activation of the emergency storage area at time t_2 and no flooding at time t_0 (the lowest branch in Fig. 2). Another scenario is to postpone the decision problem till time t_1 (the "no action" arc in Fig. 2, starting in time t_2 and ending in time t_1). In the decision tree, costs (expressed as a pair: costs of a measure and the flood damage after taking the measure) are assigned to every scenario. The values are time dependent. Indeed, increase of the time available for evacuation leads to higher evacuation costs or later activation of an emergency storage area may result in lower flood damage. The model also contains information about the flooding probability at time t_0 for a variety of circumstances. Probability P_1 is defined as a flooding probability in the region at time t_0 given information available in time t_1. Probability P_1* is defined as a flooding probability in the region at time t_0 given information available in time t_1 and knowing that the emergency storage area has been activated (in contrast to an evacuation decision, activation of an emergency storage area has an influence on the flooding probability). Probabilities P_2 and P_2* are defined analogically. These probabilities are by definition the conditional flooding probabilities. The available information can consist of water level forecasts and/or the expected condition of the dikes.

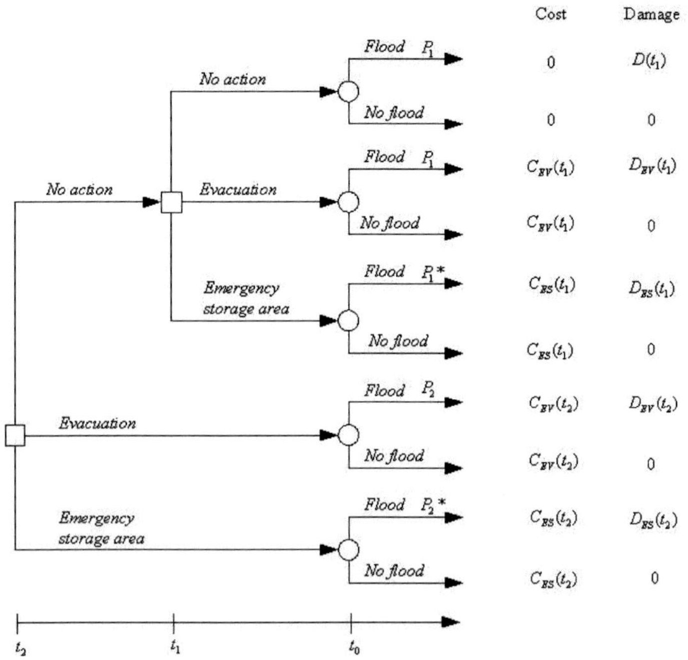

Figure 2: Application of the decision tree method.

Let us emphasize, that the decision (“no action”, “evacuation”, “activation of an emergency storage area”) has to be taken at time t_2, however, the decision tree model also requires specification of flooding probability at time t_0 given information available at time t_1 (note that time t_1 is assumed to be later than time t_2). In short, probability P_1 cannot be known at time t_2. The problem of estimating probability P_1, being at time t_2, has been tackled in the literature (see Frieser [3]). Estimation of probability P_2 can be made as shown in Wojciechowska [10]. Estimation of probabilities P_1* and P_2* has not been discussed yet.

Estimation of costs and flood damage in the case of evacuation decision is in the most part presented in Frieser [3], whereas estimation of costs and flood damage in the case of activation of an emergency storage area is a subject for future work.

3.2 Decision influence diagrams

3.2.1 Definition

A decision influence diagram constitutes a compact graphical and mathematical representation of a decision problem, leading to better understanding and consequently to (rational) solving of the problem. In general, a decision influence diagram offers an intuitive way to identify and display the important elements of a decision-making process, including decisions, uncertainties and objectives, and relations (dependencies) between them.

More precisely, a decision influence diagram is defined as an acyclic directed graph (see Barlow [11]) where the following holds:

- Circle nodes represent random variables, double circle nodes denote deterministic values and squared nodes stand for decisions;
- Directed arcs into probabilistic and deterministic nodes show possible dependence, whereas directed arcs into decision nodes signifies available information at the moment of taking decision;
- A conditional probability (deterministic) function is associated with every probabilistic (deterministic) node, whereas a collection of decision rules is associated with every decision node (for more information on this subject see Barlow [11]);
- Decision nodes are in order; a decision node is (possibly implicitly) connected to all predecessor decision nodes;
- There exists exactly one deterministic node without any successors; such node is called the value node.

There are two operations used to solve decision problems represented by decision influence diagrams. These operations are the elimination of probabilistic nodes and the elimination of decision nodes (Barlow [11]). Decision influence diagrams are derived as an alternative to decision trees and the solution process of the both representations of a decision problem is fundamentally the same (Barlow [11]).

In general, a decision tree will have many more nodes and arcs than the corresponding decision influence diagram, what can make the decision tree method quite burdensome. Decision trees are, however, more useful in presenting all outcomes of decision problems. On the other hand, decision influence diagrams emphasize dependencies and independencies among variables more clearly than the related decision tree.

3.2.2 Application

In Figure 3, a decision influence diagram is presented. The diagram corresponds to the decision problem described in paragraph 3.1.2. In other words, the diagram is an alternative to the decision tree shown in Fig. 2.

Let us consider the diagram. The diagram consists of four nodes, each with prescribed meaning. Moreover, five directed arcs are contained in the diagram. As already mentioned, the arcs indicate possible dependence. All considered decisions are represented by a one squared node. Random variability of the flood event at time t_0 is represented by a single circle node. Available information (known during the decision-making and essential for assessing the flooding probability) and values corresponding to the decision-state pairs are represented by double circles. The decisions and (uncertain) flood event have an influence on the inquired costs therefore the corresponding nodes are connected. Taken decision can have an influence on the flooding probability at time t_0, thus, a directed arc from the decision node to the random node has been drawn (e.g. activation of the emergency storage area has an influence on the flooding probability).

Clearly, the diagram has a much more compact form than the corresponding decision tree leading to better visibility of present variables and connections between them.

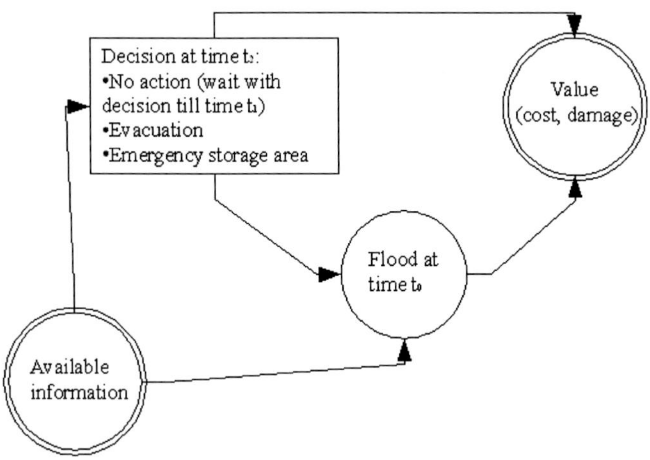

Figure 3: Application of the decision influence diagram method (alternative to Fig. 2).

3.3 Markov Decision Processes

3.3.1 Definition

Markov Decision Processes offer a mathematical structure for modelling of a decision-making in circumstances when results are partly random and partly under the influence of a decision maker.

A Markov Decision Process is a stochastic process observed at discrete time points. At every time point the process is found in one of possible states. After observing the state of the process, an action is taken. When the process is in state i at time t and action a is chosen, then (bounded) cost C(i,a) is incurred and the next state of the system, say j, is chosen with probability $P_{ij}(a)$, hence $P_{ij}(a)$ is a transition probability from state i to state j when action a is taken. Both the costs and the transition probabilities are functions only of the last state and the upcoming action. Moreover, it is assumed that the number of all actions is finite.

To choose an action (depending on a state), a decision maker must act according to some policy. A policy can be simply a function from the state space into the action space. In the case of short-term decisions (measured in days), the aim is to determine an average cost optimal policy, which minimizes average expected cost for all initial states. In general, the optimal policy can be identified using algorithms proposed in the literature (e.g. value iteration algorithm or policy improvement algorithm). For more theoretical information on the subject see Ross [12].

3.3.2 Application

Let us consider application of the Markov Decision Processes to operational flood management. For the sake of simplicity, only evacuation decision will be considered. It is assumed that a system consists of monetary consequences of (eventual) flooding and actual flooding probability. The latter is defined as a flooding probability given actual information. Further, it is assumed that the system can be found in one of five states. State 0 corresponds to an ideal situation, where flooding danger is very small (the consequences-probability pair is not specified for this state). It is assumed that state 0 is an absorbing state (it is not possible to leave this state) and that the system cannot start in this state. State 1 corresponds to the situation where the consequences of (eventual) flooding are equal to C and the actual flooding probability is equal to P_1, this is written as $1=\{C, P_1\}$. Similarly for the three following states, we define $2=\{C, P_2\}$, $3=\{C_1, P_1\}$ and $4=\{C_1, P_2\}$, where $C_1<C$ and $P_1<P_2$.

In the example two actions lead to transition between the states: evacuation and no evacuation (no action). In states 1 and 2 the two actions are possible, whereas in states 0, 3 and 4 only the latter action can be chosen.

Let us now consider the system being in state 1 at the present time point. No action can lead to state 0 (ideal situation) or to state 2 (increase of flooding probability) at the next time point, but there is also a possibility that the state of the system will not change at the following time point. Evacuation can lead to states 0, 3 or 4 at the next time point. In state 3 the consequences of flooding are smaller ($C_1<C$) and the flooding probability does not change. It is reasonable that consequences in state 3 are less severe, because evacuation reduces potential

flood damage. In state 4 the consequences of flooding are less severe, but the flooding probability increases. In states 3 and 4 no action is possible, leading to absorbing state 0. Similar considerations can be made for the system being in state 2 at the present time point, entailing Figure 4.

For a specific action, probabilities affect transitions between the states. Let us remark, that e.g. the probability of transition from state 2 to state 1, when no action is taken, is equal to the probability of transition from state 2 to state 3 when evacuation is chosen. Also, according to the principle of the method, costs are associated with every taken action. In the considered problem the costs will be additionally dependent on the following state and can be defined as the effectiveness of the chosen action (the costs will be in fact rewards). The aim of the model is to determine the best action being in state 1 or 2 (note that for the other states the "best" actions are already specified).

Let us emphasize that the presented application of Markov Decision Processes is rather fake (it is created for the purpose of this paper) and does not correspond to the classical book examples, where usually all actions are available in every state of a system (see Ross [12]). An evacuation decision can be made once; allowing the evacuation decision to be chosen for the first at time t and then again at time t+2 could lead to a paradox. The same observation holds for a Markov Decision Process, where activation of an emergency storage are is one of possible actions.

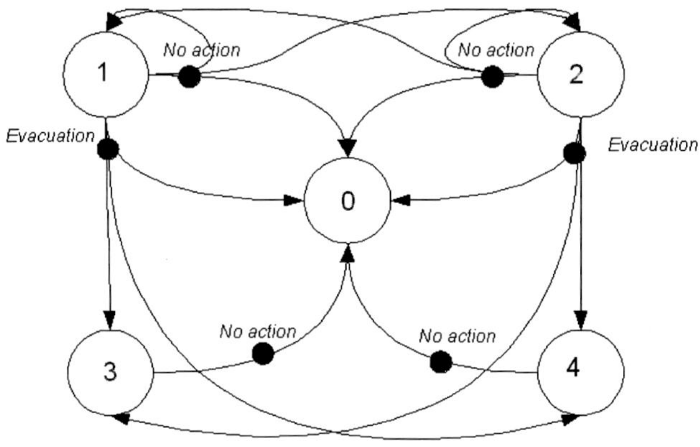

Figure 4: Application of the Markov Decision Processes.

4 Discussion, conclusions and future work

Discussion, conclusions and future work following from this research can be summarised as follows:

- As it has been already mentioned, decision trees and decision influence diagrams are alternatives – despite all differences, the solution process in both methods is fundamentally the same. We state to determine both decision tree and decision influence diagram for a decision problem, since the graphs (visually) complete each other leading to a better understanding of the problem.

- In contrast to Markov Decision Processes, the application of decision trees and decision influence diagrams is rather intuitive and does not require extensive mathematical knowledge. In general, a Markov Decision Process cannot be translated to a decision tree or a decision influence diagram – the methods tackle slightly different decision problems.
- In the presented application of the Markov Decision Processes, several parameters have to be specified (e.g. transition probabilities, probabilities P_1 and P_2). A large number of unknown parameters can constitute a difficulty during crisis situations.
- The presented application of Markov Decision Processes is rather fake and does not correspond to the classical book examples, where usually all actions are available in every possible state of a system (see Ross [12]). Evacuation decision (or, e.g., activation of an emergency storage are) can be made once; allowing an evacuation decision to be chosen for the first at time t and then again at time t+2 could lead to a paradox.
- The considered decision methods allow one to include mathematically some uncertainties. It is important to emphasize that the methods can be applied as a rational means of support in real-life situations – the methods cannot replace the usual decision-making, where often intuition or other information is used. Furthermore, the methods constitute rather a schematic approach to decision problems.
- The considered decision methods are not applied in operational flood management in the Netherlands. Introduction of the methods would certainly require time. Both developing and learning processes should be a common work of scientists and end-users (so decision makers); this could prevent a gap between the scientists' expectations and the end-users' needs.
- Future work concerning the subject should focus on the estimation of the required parameters i.e. conditional flooding probabilities P_1* and P_2*, costs and flood damage in the case of activation of an emergency storage area after (in the decision tree) or the transition probabilities (in the Markov Decision Process).

Acknowledgements

This research is part of the Flood Control 2015 program. The program is partly funded by the Dutch Ministry of Transport, Public Works and Water Management. For more information visit http://www.floodcontrol2015.com.

References

[1] Bezuyen, M.J., van Duin, M.J. & Leenders, P.H.J.A., Flood Management in The Netherlands, *Australian Journal of Emergency Management*, winter 1998.

[2] HWR, *Hoog Water 1995*, Regio Nijmegen, Evaluatierapport (in Dutch), 1995.

[3] Frieser, B.I., *Probabilistic Evacuation Decision Model for River Floods in the Netherlands*, Master's thesis, Delft University of Technology and Ministry of Transport, Public Works and Water Management, Delft, June 2004.

[4] Verkade, J., *The value of flood warning systems*, M.Sc. dissertation, Water resources Management programme Delft University of Technology, Delft, June 2008.

[5] Glassheim, E., Fear and loathing in North Dakota, *Natural Hazards Observer*, XXI, 1-4, 1997.

[6] Barendregt, A., van Noortwijk, J.M., van der Doef, M. & Holterman, S.R., Determining the time available for evacuation of a dike-ring area by expert judgement, *ISSH – Stochastic Hydraulics 2005 – 23 and 25 May 2005 – Nijmegen*, the Netherlands.

[7] RBOS, *Rampenbeheersingsstrategie Overstromingen Rijn en Maas Achtergrondrapportage Veiligheid en rivierkunde* (in Dutch), RIZA report 2005.024, Ministerie van Verkeer en Waterstaat, December 2005.

[8] Nortier, I.W. & de Koning, P., Toegepaste Vloeistofmechanica, Hydraulica voor waterbouwkundigen (in Dutch), Chapter 7. *Stroming over overlaten en meetschotten. Stuwen. Afvoercoefficient.* Wolters-Noordhoff bv Groningen/Houten, The Netherlands, pp. 278-279, 1996.

[9] Benjamin, J.R. & Cornell, C.A., Probability Statistics and Decision for Civil Engineers, Chapter 5. *Elementary Bayesian Decision Theory*, McGraw-Hill, Inc., pp. 526-581, 1970.

[10] Wojciechowska, K.A., Extension of the probabilistic evacuation decision model, *Proc. of the 7th International Probabilistic Workshop 25-26 November*, eds. P. van Gelder, D. Proske & H. Vrijling, Delft, the Netherlands, pp. 313-325, 2009.

[11] Barlow, R.E., Engineering Reliability, (Chapter 10). *Making Decisions Using Influence Diagrams*, University of California, Berkeley, the American Statistical Association and the Society for Industrial and Applied Mathematics, pp. 165-181, 1998.

[12] Ross, S.M., Applied Probability Models with Optimization Applications, (Chapter 6). *Markov Decision Processes*, Dover Publications INC., New York, pp. 119-152, 1970.

European and Chinese Integrated River Basin Management: experiences and perspectives

W. Ravesteijn[1], X. Song[2] & R. Wennersten[2]
[1]*Department of Technology Dynamics & Sustainable Development, Delft University of Technology, the Netherlands*
[2]*Department of Industrial Ecology, Royal Institute of Technology, Sweden*

Abstract

This paper addresses Integrated River Basin Management as a solution to water stress through a comparative analysis of water management in Europe and China. It makes a preliminary assessment of the European experiences so far with the Water Framework Directive (WFD) – and the additional groundwater protection and flooding directives – on river basin management and investigates its potential value for other areas, focusing on China. Problems and possibilities of water policy transfer are investigated. The main questions to be discussed are whether the EU WFD could be made part of China's water management regime and whether the experiences of the WFD could help China's water managers and decision-makers to improve water management based on the principles of IRBM. We will deal with these and other questions, discussing the differences in water management and context between Europe and China.
Keywords: China, European Union, floods, Integrated River Basin Management, Water Framework Directive, policy transfer.

1 Introduction

Water stress, including water scarcity, large-scale flooding and high-level contamination of drinking water, is a serious problem in the world of today in which regions, countries and localities experience different problems – or mixtures of problems – in different levels of severity. Flooding, e.g., threatens delta areas, while drought threatens many inland areas. Global warming aggravates these problems, including the introduction of new types of problems like flood hazards to complete delta areas, which could entail not only direct human suffering but also lead to local and international conflicts about the availability, distribution and risks of scarce water resources.

It has long been accepted that a socio-technical systems approach is critical to address water management problems and to generate solutions [1–5]. Balancing the interests of different water using sectors and different actors, combining arrangements and decisions at different scales, and linking communities to their physical resource base are seen as prerequisites for sustainable development. All of these issues can only be addressed from a systems perspective, focusing on the interactions and relations between the various components, rather than treating them as isolated issues as well as considering the dynamics of the systems themselves. Integrated River Basin Management (IRBM) – like Integrated Water Resources Management (IWRM) – clearly builds on a systems approach to water resources management and, consequently, it is a promising tool for dealing with present-day water stress.

This paper addresses IRBM through a comparative analysis of water management in Europe and China. It focuses on the European version of IRBM: the Water Framework Directive (WFD) as well as its additional groundwater protection and flooding directives and it investigates its relevance for China. Questions dealt with are: Is the WFD appropriate and feasible in the Chinese context? What would be the implications of introducing the WFD for Chinese water management? What adaptations might be necessary? How would context and adaptations influence the potential benefits of the WFD? First, we will enter into the European WFD and its backgrounds and expansions. Consequently, we will describe and analyse the water problems and policies in China. Last but not least, we will discuss the possibilities and limitations of policy transfer from Europe to China.

2 European IRBM

Water resources management and development have a long history in Europe. Trans-boundary rivers like the Rivers Rhine (1,233 km long) and Danube (2,860 km long) required conflict management for a long time, while the post-war rapid economic growth affected river systems and called for intervention too. Water-related legislation has been developed since the 1970s with various aims, e.g. reducing emissions of certain substances from different sources and water quality objectives for drinking water and bathing. During the early 1990s, however, needs emerged to address water management challenges from a more integrated and holistic perspective. In December 2000, the Water Framework Directive (WFD) was issued after long time cross-national negotiation and implemented in all 25 EU member states.

The main objectives of the WFD are:

- Expanding the scope of water protection to all waters (inland surface waters, transitional water, coastal waters and groundwater) in a holistic way
- Achieving “good status” for all waters by the target date of 2015, satisfying human needs, ecosystem functions and biodiversity protection

- Water management within the hydrological/geographical boundaries of river basins via effective cooperation of all administrations and institutions involved
- Combined approach towards the control of both point and non-point pollutant sources with emission limit values and water quality standards
- Getting the right prices of water with the elements of cost recovery and cost-effectiveness provisions
- Getting citizens more closely involved in river basin management activities
- Streamlining legislation by repelling existing fragmented and burdensome regulatory systems [6].

The European WFD provides a common framework for water policy employing integrated approaches and innovative instruments to water management. It was developed as a response to the fragmented European environmental legislation. However, the international cooperation concerning the Rhine, starting in the nineteenth century, can be considered a pioneering effort. The WFD offers a policy frame for protecting and improving the quality of water resources, though it also involves flood and groundwater quality control (for which follow-up directives were issued; see below). It provides for cooperation at the level of river basins. A precise planning procedure for river basin management is part of it (fig. 1).

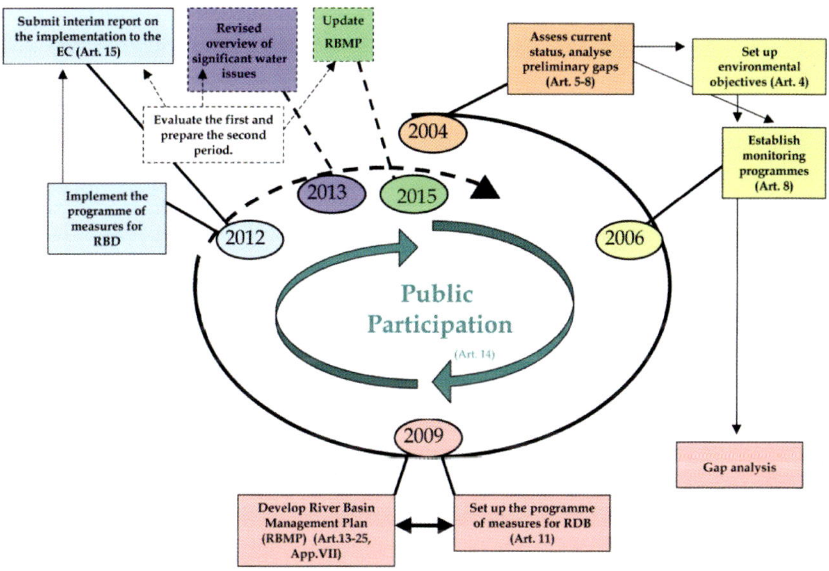

Figure 1: River basin management planning process [7].

The main point is towards water resources management at river basin level in districts. These River Basin Districts (RBD's) are largely based on surface water catchments, together with the boundaries of associated groundwater and coastal water bodies. In April 2009, the conference of "Active involvement in River

Basin Management – Plunge into the debate!" in Brussels has supported the preparation of river basin plans [8].

The WFD got a supplement in 2006 with the Directive on the protection of groundwater against pollution and deterioration [9]. In 2007, another step was taken in realizing IRBM in Europe when the European Parliament adopted the Floods Directive [10]. This requires European countries to make flood risk assessments, focused on impacts on human health and life, the environment, cultural heritage and economic activity and to do so before 2012. These assessments will be used to produce flood hazard and risk maps (planned to be ready by December 2013), on the basis of which flood risk management plans will be drawn up (to be ready by December 2015). The latter plans will be developed together with all relevant stakeholders and should provide policy makers and developers with a base for measures that are both technically appropriate and socially acceptable.

The (expanded) WFD is the European edition of IRBM, stimulated and being conditioned by the post-war European unification process, in which public control measures follow on the establishment of a free market. The concept of IRBM comprises multi-purpose water projects, basin wide programs and comprehensive regional development. It has been created in the United States of America in the first half of the twentieth century on the basis of engineering and management experiences from all continents [4]. In fact, IRBM being a management tool strongly reflects the doctrines of planning, which originated from the Second Industrial Revolution taking place in Germany and the USA. Planning procedures were part of Roosevelt's New Deal politics (e.g. the Tennessee Valley Authority), but became a dominant policy instrument first in the Soviet Union. After the Second World War, planning as part of management became widely adopted. Similarly, IRBM has been broadly accepted as a framework for water policy, in which water developers often combine modern concepts with endogenous water traditions. The European WFD is a case in point.

The EU WFD and its supplements, tailored to the specifics of the European countries, reflect many but not all the elements of IRBM. Experiences thus far are positive. Though it is causing tensions here and there in view of national water traditions [11], in general, the WFD seems to work quite well; the goal at least – in 2015 all waters in a "good condition" – is still within reach. However, recent assessments point out that at least 40% of the European water bodies are "at risk" not meeting the goal [12]. Nevertheless, the WFD has been hailed as a front-runner on integrated water management in the world owing to the introduction of a number of generally agreed principles and concepts into a binding regulatory instrument [13].

Consequently, the WFD and European IRBM in general have become a source of inspiration for water management reforms elsewhere, most notably in China. The Chinese water authorities have shown interests in adopting the WFD to fight the river pollution problems and manage flood risks that they are experiencing, e.g. in the cases of the Yellow River and the Yangtze River.

3 Chinese water management

Since the 1980s, China has experienced different kinds of water stress, especially floods, droughts, water pollution and the degradation of terrestrial and aquatic ecosystems, mainly due to the economy-orientated development mode, the huge population growth and the accompanying altered land use. Historically, China has been frequently hit by major floods and it has suffered from flooding disasters as well as droughts. In recent decades, China has put much effort on flood control, but the current challenges are still gigantic. During the period 2001-2005, the average area of land affected by floods was 12.8 million ha, with 1,510 people dead and a direct economic loss of 100.6 billion Chinese Yuan [14].

River pollution becomes more and more serious. The annual quantity of wastewater discharged increased from 58.4 billion m^3 in 1997 to 75.8 billion m^3 in 2008, of which around two-thirds was industrial wastewater and one-third domestic sewage [15]. Diffuse emissions from agricultural activities contribute to excessive loads of nutrients in rivers and lakes. In 2008, 20.6% of all 150,000 km monitored river sections had a water quality below Grade V (essentially useless), and 18.2% was Grade IV–V (appropriate only for agriculture and industrial use) [15].

At the same time, the competition for water among agriculture, industry and municipalities is becoming more and more intense as a result of rapid urbanization, industrialization and sustained population growth. The total quantity of water used in 2008 was 591 billion m^3, of which 12.3% was for domestic water use, 23.7% for industry, 62% for agriculture and only 2% for the environment [15]. Moreover, various trans-jurisdictional conflicts are emerging, e.g. on water rights and pollutants discharge.

3.1 Laws

China has developed a set of water relevant laws and regulations since the 1980s, including the Prevention and Control of Water Pollution Law (1984, amended in 1996 and 2008), the Regulations for River Administration (1988), the Environmental Protection Law (1989) and the Water and Soil Conservation Law (1991, amended in 2010). The Flood Control Law, adopted in 1997, defines various mechanisms and measures for flood control planning and management. The main law, however, is the Water Law, initiated in 1988 and amended in 2002.

China's water resources have been managed by means of administrative boundaries for decades. This pattern of administrative water management was legally strengthened by the first Water Law adopted in 1988, which emphasised that "the State shall exercise a system of unified administration on water resources in association with administration at various levels and by various departments" (Art. 9). It marks China's water management shift from "administrative management" to "water governance by law". However, several major rivers run through many administrative regions in China. The Yangtze

River (6,300 km long), for example, flows through 11 provinces including autonomous regions. Similarly, the Yellow River (5,465 km in length) runs through nine provinces and autonomous regions.

The River Basin Commissions (RBCs) actually started to develop since the 1930s; their main functions were water quantity management – e.g. aiming at flood prevention and keeping rivers running – and water and soil conservation. Consequently, there have been various upstream-downstream conflicts about water allocation, water pollution and river development, which the first Water Law (1988) failed to deal with.

Since the promulgation of the amended Water Law in 2002, China's water management has started to move towards a new system using both administrative and hydrological boundaries. This is emphasised in several ways, such as "The State shall adopt a 'combined division responsibility' system of river basin management in conjunction with jurisdictional management" (Art. 12) and "the plan for a region within a river basin shall be subordinated to the comprehensive river basin plan" (Art. 15).

The amended Water Law also indicated that "The River Basin Commissions shall perform the duties of water resources management and supervision, which are specified by laws and administrative regulations, and granted by water administrative departments under the State Council; The administrative water management departments above the county level shall be responsible for the unified water resources management and supervisions in their regions" (Art. 12). Moreover, it emphasised that "Water conflicts among administrative regions in river basins shall first be negotiated; if it fails, the case shall be sent to a higher level of government for judgment" (Art. 56).

3.2 Institutional structure

China has a long history of river basin development and management in various forms, including the establishment of seven river basin commissions (RBCs) since the 1930s and consequent attempts to develop river basin plans. RBCs primarily focused on traditional hydropower and flood control projects. Until now, they hardly apply the principles of Integrated River Basin Management, while their institutional capacity for water conflict resolution among different administrative regions is weak. Water quantity management is emphasized, rather than water quality management. Furthermore, there are many unclarities in the division of water resources management between the administrative agencies and the RBCs. In practice, this may slow down action in emergency situations, such as large-scale water pollution and flood events in river basins, when fast and effective responses are required.

China has a highly top-down water administrative structure, involving a variety of water-related organisations. At the national level, there are a couple of departments administering water resources affiliated to the State Council, e.g. the Ministry of Water Resources (MWR), the Ministry of Environmental Protection (MEP) and the State Flood Control and Drought Relief Headquarters (SFDH). In this context, water institutional arrangements are often criticised for being inefficient, sectoral and incapable [16, 17].

An example of the current fragmented water management system is the separate water quantity and water quality management framework (fig. 2). At the river basin level, the MWR is mainly responsible for water quantity management in water systems, while the MEP is mainly responsible for pollutant discharges from land to water. The Water Protection Bureau of the RBCs, assisting in water quality management, is led by the MEP as well as the MWR. Consequently, it is very hard to coordinate water quantity and quality management in river basins. Actually, the core of the present river basin management is to control floods and to coordinate water allocation in view of different demands. In order to improve the practices of IRBM in China, it is very important to first clarify the various responsibilities of water institutions at all levels and then develop effective coordination mechanisms suitable for China's national conditions.

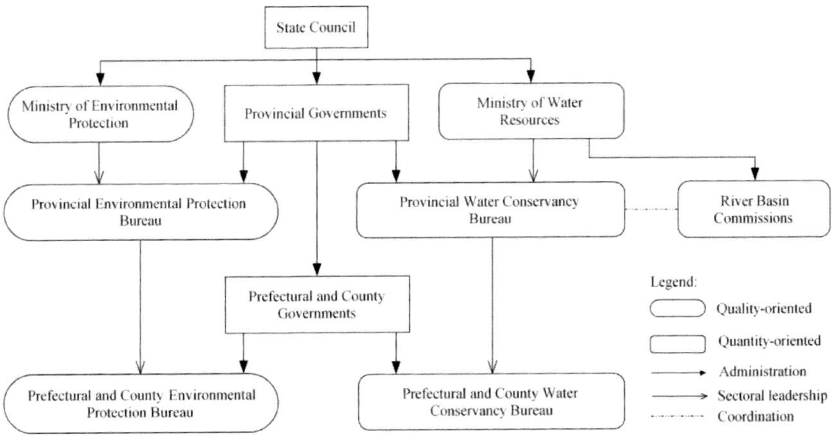

Figure 2: Separate water quality and quantity management in China.

3.3 IRBM

Though the concepts of sustainable development are gradually becoming prominent in China, water management policies and practices towards IRBM (and IWRM in general) remain out-of-date and fragmented. The awareness grows, however, that comprehensive water resources development and management at river basin level is crucial to improving the current water situation and to ensuring the long-term availability of water resources as a base of social welfare [18].

Problems to overcome include weak water governance, low level of public participation, inefficient implementation of laws and regulations and overlapping legal and institutional functions. IRBM-based water reforms are necessary, and should include strengthening and improving the legal position of RBCs, water demand management, flood risk management, pollution control and prevention, balanced hydro-ecological system protection and economic development, coordination with regard to laws, regulations and institutions as well as appropriate planning and decision-making procedures. International experiences, e.g. with the EU WFD, could help China in this regard.

4 Discussion

IRBM has attracted attention from both developed and developing countries in addressing present-day water stress. Valuable lessons can be learned from experiences elsewhere, for example in Europe. However, policy transfer is far from easy. Hooper [19, 20] argues that south-south exchange of experiences is more likely to promote effective IRBM implementation than north-south dialogue, partly because of divergent basin realities. Attempts to use the Murray-Darling model in Vietnam and China have encountered serious problems [21]. These cases show that river basin management strategies must be adapted to local situations and that institutional water reforms in developing countries cannot result from duplicating management in developed countries. Does this also hold for implementing European IRBM in China? Could the EU WFD be adopted in China, having the same beneficial effects as in Europe?

China is facing major challenges in managing its scarce water resources to sustain economic growth and it makes tremendous attempts to fight water stress with elements from foreign water approaches, including IWRM methods. Some researches and activities have been carried out in order to find and assess learning possibilities, involving e.g. water management in Austria [22], the EU and USA [23] as well as the International Yellow and Yangtze River Forum. The WFD inspired the EU-China River Basin Management Programme (2007-2012, focussed on the Yellow River and Yangtze River basins [24]).

In order to answer the above questions, there are some facets to take into consideration, including differences and similarities between China and Europe and the water situation in both regions [25]. Table 1 gives an overview of differences in water management in the EU and China, reflecting differences as to the specific water stress situation, both physically and perceived, development status, policy priorities as well as the broader socio-economic, political-economic and socio-cultural context and historical backgrounds.

The differences in the present-day water regimes as well as in the social, economic, political and cultural context between the EU and China seem to preclude the complete importation of the WFD. Three points deserve attention.

1. A striking difference is the political structure as well as political tendencies, which determine developments and possibilities in the water domain. Presently in China, central guidance in water management to address the varied and complex water problems is getting more and more difficult, which implies the necessity of exploring combined central and decentral strategies to be explored. The European case offers an interesting example in this respect, though the other way around: Europe is moving toward increased central guidance in a basically decentralized (nationally organized) social system, whereas China tries to adopt decentral elements in its water management approaches in order to solve complex water administrative problems. China and Europe are

representing two archetypal organizational models for implementing IRBM, i.e. the authority model and the coordination and negotiation model. Interestingly, however, a tendency to convergence can be noted, though great differences still remain.

2. It makes sense to manage rivers in regard to hydrological boundaries. Consequently, river basin administrations are considered to be of essential importance in both China and Europe. The problem, however, is how to distribute power between China's administrative water management agencies and RBCs. China definitely needs rivers to be fully managed and developed in terms of complete basins as Europe presently strives after, but the question remains how to adapt the EU's river basin management framework to China's specific socio-ecological context.
3. The involvement of local stakeholders in basin-level planning and actions was right from the start a main point in Europe, logically related to the political organization of decision-making and policy implementation in Europe. The April 2009 Active involvement in River Basin Management conference in Brussels exemplifies this. It brought decision-makers and stakeholders from all across Europe together in view of the formulation and publication of river basin plans before 2010 [7]. How to realize such in present-day China? In general, China's progress in realizing IRBM depends on the public awareness of environmental problems in relation to economic growth and the development of a civil society.

Table 1: Comparison of contemporary water management between Europe and China.

Aspect	EU (WFD)	China
Objectives	Good status of all waters(including surface and groundwater)	Water conservation and pollution prevention (mainly surface water)
Scope of planning	River basin planning, update river basin plan every six years	Combination of river basin planning and regional administrative planning
Pollutants management	Combined approach towards control of both point and non-point source pollutants	Focus on the control of point source pollutants; no effective measures for non-point source pollutants
Decision making	Top-down and bottom-up, centralized and decentralized management	Top-down, centralized management with strong and varied hierarchy
Role of RBCs	Responsible for all water-related issues at river basin level	Mainly responsible for water quantity management, e.g. flood control and water allocation
Water allocation and water rights	Controls on water abstraction and groundwater recharge; member states' own policies specify water rights	Rational allocation to alleviate the upstream-downstream conflicts; ambiguous water rights at regional and local levels
Water pricing	Full cost recovery and cost-effective provisions to be taken into account	Preliminary research on water price, its components, and measurement
Public participation	Getting citizens more closely involved in river basin management activities from early stages	Insufficient stakeholder participation in water planning and management as well as flood control

China's legislation on IRBM does not provide clear guidance on how to achieve it. European policies can be used as an example. Theoretically, the European approach is helpful for China, especially with regard to alleviating river pollution and ecological degradation. However, the contextual differences on hydrology, water stress, legal and institutional structure and the like clearly indicate that WFD cannot be applied directly to China.

5 Conclusion

The experiences with the EU WFD and European IRBM in general are far-reaching and relevant beyond the scope of Europe. It changes the widely held view that water resources primarily serve economic development. Furthermore, it puts forward a series of management facets, e.g. clear and effective water legislation, public participation, water management on hydrological boundaries and cost recovery, which offer potential lessons for the development of integrated river management elsewhere. Consequently, the WFD – as an example of IRBM – has attracted foreign attention, most notably from the Chinese water authorities.

Controlling rivers has been the main priority in China's traditional "hydraulic society", especially involving dam construction and long distance water diversion. Although water management strategies have shifted from "water engineering" to an approach of managing and developing "water resources" in the 1990s, more adaptive strategies towards IWRM and IRBM in China are necessary for fighting the massive river pollution, the frequent seasonal floods and other water problems at the scale of a river basin.

However, there are large differences in the scope and scale of water management as well as in the geo-morphological, social, economic and cultural context between Europe and China, which seem to preclude the complete importation or 'transplantation' of the WFD and its supplementary directives. Furthermore, European IRBM is very complex and requires a lot of (inter)national cooperation. The implication for China is that pure legislative and institutional reforms do not suffice. The water management agencies should get full opportunity to supervise integrated water resources planning and management, enforce water laws effectively and promote public participation at all different levels.

Nevertheless, the experiences with the WFD and European IRBM in general can be valuable for river basin management in China. Given all the water management aspects it introduced and systemized, the WFD could serve as a reference case. It could facilitate the development of more effective river basin management systems in China in the long term, supporting the country's rapid processes of social and economic change and transformation, while taking into account the particularities of individual river basins.

References

[1] Ravesteijn, W., Hermans, L. & Vleuten, E. van der, Water systems: Participation and globalisation in water system building. *Knowledge, Technology & Policy* **14(4)**, pp. 4-12, 2002 (special issue).

[2] Bressers, H. & Kuks, S. (eds.), *Integrated Governance and Water Basin Management*, Kluwer Academic Publishers: Dordrecht/Boston/London, 2004.

[3] Kaijser, A., System building from below: Institutional change in Dutch water control systems. *Technology and Culture* **43(3)**, pp. 521-548, 2002.

[4] Kates, R.W & Burton, I. (eds.), *Geography, Resources and Environment. Vol. 2: Themes from the Work of Gilbert F. White*, The University of Chicago Press: Chicago, 1986.

[5] Hoekstra, A.Y. & Huynen, M., Balancing the world water demand and supply (Chapter 2). *Transitions in a Globalising World*, eds. P. Martens & J. Rotmans, Swets & Zeitlinger Publishers: Lisse, pp. 17-35, 2002.

[6] European Commission (EC), Introduction to the new EU Water Framework Directive, http://ec.europa.eu/environment/water/water-framework/overview.html, 2000.

[7] Europe Environmental Agency (EEA), River basin management plans and programme of measures, http://www.eea.europa.eu/themes/water /water-management/themes/water-management/river-basin-management-plans-and-programme-of-measures (accessed on April 11, 2009).

[8] European Water Conference (EWC), http://ec.europa.eu/environment /water /index_en.htm, 2009.

[9] European Commission (EC), Directive 2006/118/EC of the European parliament and of the Council of 12 December 2006 – on the protection of groundwater against pollution and deterioration, http://www.central 2013.eu/fileadmin/user_upload/Downloads/Document_Centre/OP_Resou rces/08_Directive_2006_118_EC.pdf, 2006.

[10] European Commission (EC), Directive 2007/60/EC of the European parliament and of the Council of 23 October 2007 – on the assessment and management of flood risks, http://eur-lex.europa.eu/LexUriServ /LexUriServ.do?uri=OJ:L:2007:288:0027:0034:EN:PDF, 2007.

[11] Ravesteijn, W. & Kroesen, O., Tensions in water management: Dutch tradition and European policy. *Water Science & Technology* **56(4)**, pp.105–111, 2007.

[12] European Commission (DG Environment), Water note no. 2, Wise 2008, ec.europa.eu/environment/water/water-framework/pdf/water_note2 _cleaning_up.pdf, 2008.

[13] Commission of the European Communities (CEC), Towards Sustainable Water Management in the EU; First stage in the implementation of the Water Framework Directive 2000/60/EC, Brussels, 2007.

[14] Wang, S. Taking the Scientific Concept of Development as Guideline and Promote Water Resources Development in the 11th Five-Year Plan Period

in an All-round Way. In Zhou, Y. (ed.) *China Water Development Report 2006* (in Chinese). China WaterPower Press, Beijing, pp. 3-21, 2006.

[15] Ministry of Water Resources (MWR), China Water Resources Bulletin (in Chinese), http://www.mwr.gov.cn/zwzc/hygb/szygb/qgszygb/201001/t20100119_171051.html, 2008.

[16] Varis, O. & Vakkilainen, P., China's 8 challenges to water resources management in the first quarter of 21st Century. *Geomorphology*, **41(2-3)**, pp. 93-104, 2001.

[17] Yan, F., He, D.M. & Beth, K., Water resources administration institutions in China. *Water Policy*, **8(4)**, pp. 291-301, 2006.

[18] Song, X, Ravesteijn, W., Frostell, B. & Wennersten, R. Managing water resources for sustainable development: The case of integrated river basin management in China. *Water Science &Technology*, **61(2)**, 499-506, 2010.

[19] Hooper, B., *Integrated River Basin Governance: Learning from International Experience*. IWA, London, 2005.

[20] Hooper, B., Integrated water resources management: Governance, best practice, and research challenges. *Journal of Contemporary Water Research & Education* **135**, pp. 1-7, 2006.

[21] Svendsen, M., Integrated management of water in river basins (Chapter 13). *Overcoming Water Scarcity and Quality Constrains*, eds. R.S. Meinzen-Dick & M.W. Rosegrant, 2020 Vision Focus No. 9, International Food Policy Research Institute, Washington D.C., pp. 26-27, 2001.

[22] Hu, X., Integrated catchment management in China: Application of Australian experiences. *Water International*, **24(4)**, pp. 293-306, 1999.

[23] World Bank, Evolution of Integrated Approaches to Water Resource Management in Europe and the United States. World Bank Analytical and Advisory Assistant Programme, Background Paper No.2, 43 pp., 2006.

[24] Wang, Y., Li, L., Wang, X., Yu, X. and Wang, Y., *Taking Stock of Integrated River Basin Management in China*, Science Press: Beijing, 2007.

[25] Jong, M.D., Lalenis, K. & Mamadouh, V., *The Theory and Practice of Institutional Transplantation: Experiences with the Transfer of Policy Institutions*, Kluwer Academic Publishers, The Netherlands, 2002.

On the advantage of using risk curves to assess flood protection measures

B. Dewals, S. Detrembleur, P. Archambeau,
S. Erpicum & M. Pirotton
Hydrology, Applied Hydrodynamics and Hydraulic Constructions (HACH), Department ArGEnCo, University of Liege, Belgium

Abstract

Modern flood management approaches require the quantification of flood risk, accounting for the hazard component (flood frequency and inundation intensity) as well as the vulnerability of the floodplains (exposure, value and susceptibility). In this chapter, we present a detailed flood risk model, in which flow computation, monetary valuation of the assets and damage calculation are conducted at the scale of individual buildings or facilities. To avoid the shortcoming of focusing on economic damage, psycho-social impacts of floods are also included in the analysis. The model has been applied to evaluate three flood protection measures on a river reach in the Meuse basin (Belgium). The resulting risk curves show that such a micro-scale risk analysis provides important insights into the relative influence of the different flood protection measures. This could neither be evaluated through a more standard hydraulic analysis nor through the quantification of flood risk by only a single number.
Keywords: flood risk model, micro-scale, psycho-social impacts, detailed hydraulic model, finite volume.

1 Introduction

In Europe, flood management is gradually shifting from conventional flood defence towards more integrated flood risk management, accounting for both the hydrological/hydraulic component and the potential consequences. This trend is supported by the EU Floods Directive 2007/60/EC (EU 2007) [8], which requires flood risk management plans to be prepared by member states until 2015.

Cost-benefit analysis is explicitly mentioned in the Directive as a tool to evaluate protection measures, particularly those having a transnational effect. In a cost-benefit analysis approach, implementation of a measure should be prioritized if its cost is exceeded by the resulting benefits, which may be evaluated as the annual avoided risk cumulated over the lifetime of the measure (WG F [14]).

Compared to the conventional design of flood defences based on a single design discharge or return period, assessing measures within a risk management approach requires a far more comprehensive analysis of the various components of flood risk, including hazard and vulnerability.

Flood hazard is given by the relationship between the exceedance frequency of a flood event and the corresponding inundation intensity. The extent of flooded area and the water depth in the floodplains are usually considered as the most influencing inundation characteristics. The vulnerability component of risk quantification depends on two factors:

- exposure: which assets are located in the affected floodplains?
- susceptibility of the assets: to which extent do these assets encounter damage for a given inundation intensity?

In this chapter, we provide new contributions to the assessment of the annual avoided risk as a result of the implementation of protection measures. We particularly show that expressing the risk by a single number, such as the expectation value of annual risk, may be advantageously complemented by an analysis of the whole *risk curve*, which gives the amount of damage as a function of the flood frequency for a wide range of events (Kaplan and Garrick [9]).

We use a self-developed detailed 2D hydraulic model for evaluating the inundation characteristics (Erpicum *et al.* [6]), while the flood frequency is deduced from hydrological statistics. Using high resolution topographic data, the hydraulic model succeeds in accurately simulating the flows in each street and around each building, even in complex urbanized floodplains.

The vulnerability analysis is conducted here at the micro-scale, considering as the unit of analysis the individual buildings and facilities, characterized by their own nature and value. This differs from most existing studies, in which the unit of analysis corresponds to areas of a given landuse category, characterized by a specific value (per unit area).

A common shortcoming of many flood risk models is the poor consideration of intangible flood impacts, like social impacts; resulting in an inappropriate allocation of resources and consequently a suboptimal implementation of flood protection measures. Therefore, the present model includes a psycho-social impact analysis, which combines the flood hazard characteristics with social vulnerability aspects such as susceptibility of people and adaptive capacity of the community (Ernst *et al.* [5]).

The resulting micro-scale flood risk model has been tested on a case study situated along river Ourthe, one of the main tributaries of river Meuse in Belgium. The effectiveness of three structural protection measures has been assessed through the analysis of their influence on the risk curves.

In section 2, we present the risk model. The case study is presented in section 3, and the results are discussed in section 4.

2 Flood risk model

Sound flood risk analysis requires an overall consistency throughout the modelling, between spatial resolution, considered processes and accuracy; particularly as regards hazard and vulnerability modelling, as well as available input data and expected results (Apel *et al.* [1]). The analysis presented here relies on a micro-scale procedure approach, which involves hazard modelling by means of detailed 2D inundation modelling and modelling of socio-economic vulnerability based on high resolution land use, cadastral and statistical database.

2.1 Hazard modelling

Flood hazard is the relationship between flood frequency and the corresponding intensity of inundation, which may be described by numerous hydrological and hydraulic parameters such as inundation extent, water depth, flow velocity, water rising rate or flood duration.

Flood frequency may be deduced either from measured time series at a gauging station, or from results of continuous rainfall-runoff simulations.

Inundation modelling has been conducted so far based on models characterized by very different levels of complexity. These differences result from the spatial description of the main riverbed and floodplains (1D cross-sectional averaged, 2D depth-averaged, linked 1D-2D models ...), as well as from possible simplifications in the flow model (kinematic, diffusive or dynamic waves).

Here, we use the detailed 2D flow model WOLF developed at the University of Liege (Erpicum *et al.* [6]). Based on the fully dynamic wave equations, it includes no simplifying assumption in the mathematical model; apart from those resulting from depth-averaging (low vertical velocity compared to the horizontal one, leading to a hydrostatic pressure distribution across the water depth).

Such two-dimensional model represents reliably the hydrodynamic interactions between the main channel and the floodplains. It also provides the distribution of water depth and flow velocity in the floodplains, so that both the static and dynamic impacts of the flow may be characterized for all considered assets, such as individual buildings. The modelling approach also applies for extreme floods such as induced by dam and dike break or dam breaching (Dewals *et al.* [4]; Erpicum *et al.* [7]; Roger *et al.* [11]).

The model has been run on a highly accurate Digital Surface Model (DSM) resulting from the combination of laser altimetry and, wherever available, sonar bathymetry. The typical grid spacing for the simulations is kept as low as 2m, which is definitely fine enough to represent the complex flow patterns occurring at the scale of individual buildings and streets in urbanized floodplains (Ernst *et al.* [5]; Erpicum *et al.* [6]).

This model has been extensively applied since 2003 to issue inundation maps throughout the Walloon region based on 2D hydraulic modelling of over 1,000 km of rivers, for which validation has been systematically conducted (Erpicum *et al.* [6]). In addition, the selection of such a modelling approach is consistent with the subsequent risk analysis conducted at the micro-scale.

2.2 Vulnerability modelling

The outcomes of the detailed inundation modelling constitute suitable inputs for the micro-scale exposure analysis, performed using detailed land use maps and geographic database, as well as the subsequent evaluation of direct economic damage and social impacts of floods.

2.2.1 Exposure

The individual buildings and facilities constitute the unit of analysis in our micro-scale risk model. Their exposure has been determined from the combination of the inundation extent with high resolution land use database and complementary geo-referenced datasets.

In the area of the case study treated hereafter, two accurate land use vector database were available: *Top10v-GIS* and *PICC*, respectively provided by the Belgian *National Geographic Institute* (IGN) and the *Service Public de Wallonie* (SPW). The former contains 18 layers of information, such as land use, structures or hydrographical network, while the later is based on stereoscopic aerial imagery restitution and a manual post-processing consisting in enhancements such as house numbering and street names. They complement each other to achieve an optimal identification of individual buildings (geometric feature) and of their type (semantic feature).

Assignment of inundation characteristics to individual buildings is not straightforward based on the results of the detailed 2D hydraulic model. Indeed, computed water depths and other flow parameters inside the buildings are zero since all these obstacles are incorporated in the Digital Surface Model used for flow simulation. Therefore, a neighbourhood analysis has been conducted for each asset, in order to assign flow properties based, for instance, on the average of computed values in the surrounding cells (Figure 1).

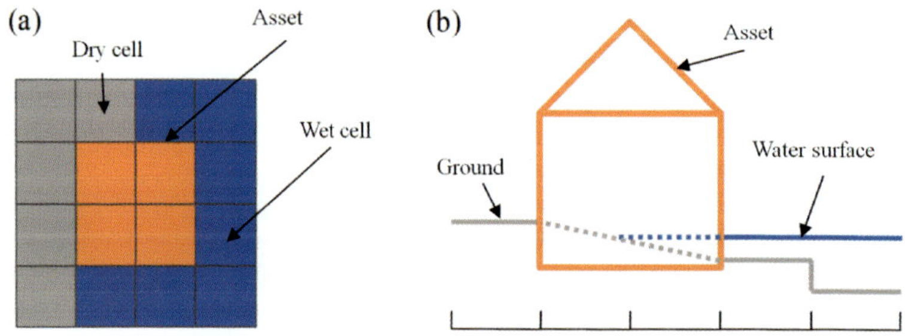

Figure 1: Neighbourhood analysis to determine the exposure of buildings: (a) plane view; (b) elevation.

2.2.2 Economic damage

A standard approach for assessing economic damage is the use of damage functions. They establish the link between the damage induced to a particular type of property and the hydraulic parameters. They usually only account for water depths, although other parameters also influence damage, such as flow velocity, precautionary measures, duration of flooding or water contamination.

Since our objective was not to develop new damage models, we used the state-of-the-art FLEMO model (Flood Loss Estimation Model) developed by Kreibich *et al.* [10] and Thieken *et al.* [13]; as well as its latest development referred to as "FLEMOps" (Flood Loss Estimation Model for The Private Sector). This model applies at the micro-scale (Apel *et al.* [1]) and expresses the relative damage as a percentage of the value of the assets; enabling thus to take into consideration local estimations of the property values.

In Belgium, the *Land registry* geographic database lists all private properties with their cadastral income, which constitutes a valuable input to estimate the economic value of the assets. Although geo-referenced, the land registry is characterized by a significantly lower geometric accuracy compared to *Top10v-GIS* and *PICC*. Therefore, land registry data have only been exploited for their semantic contain. The economic analysis presently focuses on direct potential damage to housing, reported as the main contribution to the overall flood damage in the case study area considered below.

2.2.3 Social impact

Three main drivers are considered to quantify social impacts by means of a composite index (Ernst *et al.* [5]): the flood characteristics, the susceptibility of people at risk and the adaptive capacity of communities (Coninx *et al.* [3]).

- First, local water depth, flow velocity, water rising rate and flood duration are combined to calculate a flood index (*F*), in which the relative weight of each factor is based on empirical knowledge of the correlations between these flood characteristics and the social flood impacts.
- Next, six indicators of susceptibility are calculated from the ratios of elderly people in poor health, single parent families, foreigners, financially-deprived people and people living in single-storey houses, compared to the total population (Tapsell *et al.* [12]). Such socio-economic information is available in Belgium at the level of the statistical district, which covers typically 0.25 to 1 km^2. The latest Belgian national socio-economic population survey dates back to 2001. These six indicators are aggregated by geometric mean and identical weighting factors, into a single susceptibility index (*S*).
- Third, a new analytical methodology was recently developed by Coninx and Bachus [2] to quantify the adaptive capacity of communities at the level of Belgian municipalities. The analysis is based on several factors, such as existing risk communication or non-domestic protection measures, to derive an adaptive capacity index (*A*).

These three factors are then combined as follows to evaluate the social impact index (SI): $SI = F^{\alpha} \times S^{\beta} \times A^{\gamma}$. The weights α, β, and γ have been derived from a Delphi study.

The index SI takes values between zero and one, referring respectively to limited and severe social flood impact intensity. The social impact of flooding can therefore be expressed as the number of affected people within different social impact categories (e.g., low: $0 < SI \leq 0.33$, medium: $0.33 < SI \leq 0.66$ and high: $0.66 < SI$).

2.3 Expressing risk

Following Kaplan and Garrick [9], a general expression of risk is the so-called risk curve, providing the relationship between flood frequency and potential damage for a wide range of flood intensity. In practice, the risk curve is given by a set of N pairs (f_i, D_i), with D_i the potential damage or impact (Euros, number of affected people...), f_i the corresponding exceedance frequency (inverse of the return period) and N the number of considered flood events. The annual expectation value of damage $\langle D \rangle$ is obtained by integrating the risk curve:

$$\langle D \rangle = \sum_{i=1}^{N} \left[\left(f_{i-1} - f_i \right) \frac{D_{i-1} + D_i}{2} \right],$$

where the composite trapezoidal rule has been used. The subsequent analysis relies both on the annual expectation value of damage and the complete risk curve, highlighting the complementary nature of those two expressions of risk.

3 Case study

The case study focuses on three reaches of river Ourthe in the Meuse basin. They are located a few kilometres upstream of the mouth of river Ourthe into river Meuse and have a catchment area of 2900 km². Floodplains are mostly urbanized and used for residential, commercial, touristic or industrial purposes. The 100-year discharge is estimated at 876 m³/s. Recent severe floods occurred in 1993 (742 m³/s), 1995 (520 m³/s), 2002 (570 m³/s) and 2003 (508 m³/s).

3.1 Hydrological and hydraulic characteristics

Discharge values for a number of return periods were obtained from a standard statistical analysis of a 30-year long time series of measured flow rates at a nearby gauging station along river Ourthe, as detailed in Ernst *et al.* [5].

In the case study area, the valley of the river is relatively narrow and steep. Consequently, flood waves along this type of rivers are hardly attenuated and the inundation characteristics may reliably be deduced from steady flow simulation (Erpicum *et al.* [6]). Thanks to this assumption, sequential meshing could be used to speed up computation on the fine computational grid of the model (2m × 2m). Fourteen discharge values were used as upstream boundary conditions, leading to a corresponding number of danger maps detailing the inundation characteristics as a function of the exceedance frequency.

3.2 Considered protection measures

The risk model has been applied to evaluate three flood protection measures:

- increase in the available cross-section of the river by rehabilitating an old canal formerly exploited for navigation;
- increase in the conveyance capacity of a floodplain, leading to a decrease in water head upstream;
- heightening of existing protection walls by means of mobile dikes.

4 Results and analysis

The influence of each flood protection measure has been analyzed in terms of flood hazard (inundation characteristics), psycho-social impacts and potential economic damage.

4.1 Rehabilitation of an old canal

The rehabilitation of the old canal leads to a decrease in water depth nearby the inlet of the canal and a local increase in the vicinity of its outlet into the main course of the river. The later effect is due to the lower flow velocities in this area (larger cross-section) while the water head remains essentially unchanged. This increase in water depth is located in a non-urbanized area (meadows). Hazard is thus slightly increased in a low-vulnerability area.

Implementing this measure is found to have the most significant effect on inundation characteristics only in a limited range of discharge:

- For discharge values up to the 100-year flood, an existing protection wall along this reach of the river is not overtopped both in the base scenario and in the scenario with the canal rehabilitated.
- In contrast, for flood events in the range 100-year to approximately 200-year, the decrease in water depth resulting from the rehabilitation of the canal enables to prevent overtopping of the protection wall, while this overtopping occurs in the base scenario. Preventing the overtopping leads to a decrease by about 1 m in the water depth in the protected floodplain, which is densely inhabited. Consequently, in this range of discharge, the overall effect of the measure is to shift flood hazard from a high-vulnerability area (inhabited floodplain) towards a low-vulnerability area (meadows near the outlet of the canal).
- For more extreme flood events, the existing protection wall is overtopped whatever the implementation or not of the measure. Still, rehabilitating the old canal enables to reduce the water depth in the floodplains by up to 30 cm thanks to the canal acting as a derivation and thus increasing the local cross-sectional area of the river.

4.2 Recalibration of a floodplain to increase the conveyance capacity

Compared to the base scenario, increasing the conveyance capacity of another floodplain by recalibrating its topography enables significant flow velocities to develop over a wider part of the cross section of the river and, therefore, reduces the overall flow resistance. This results in a decrease in water depth in the centre of a village located upstream of the recalibrated floodplain. This decrease is of the order of 50 cm for the 100-year flood. This hydraulic effect has been verified to act effectively for the whole range of considered flood events (return period of up to 1140 years). For the most extreme event considered, hydraulic modelling has revealed that the measure enables also to prevent inundation of a railway and a large inhabited area.

4.3 Heightening of existing protection walls (dikes)

As disclosed by the hydraulic simulations, heightening of existing protection walls along another reach of the river may lead to inundation alleviation, no effect or a worsening of the inundation characteristics depending on the type of flood event. The centre of a town is located in the corresponding floodplains.

- For flood events with a return period below 20 years, heightening the protection walls has no influence since they are not overtopped in either case.
- For flood events with a return period in-between 20 years and about 100 years, the walls get overtopped without mobile dikes whereas this overtopping is prevented by the considered heightening of the walls. In this range of discharge values, implementing the measure would thus significantly contribute to a decrease in inundation intensity in the centre of the town.
- For higher discharges, the hydraulic modelling shows that the inundations characteristics in the town get worse in the scenario with heightening of the walls compared to the base scenario. Indeed, for this range of discharge, the heightened walls get also overtopped, while the presence of the heightening induces an additional flow resistance and therefore higher water depths.

Consequently, this measure is found to have the capacity of alleviating inundations up to some design discharge (about 100-year flood); but also to intensify the danger in the case of more extreme flood events.

4.4 Effect of protection measures on flood risk

Figure 2 shows the risk curves obtained in the base scenario and accounting for the three considered protection measures. Similar figures may be drawn to express the psycho-social risk (number of affected people for different social impact intensities vs. flood frequency). The risk curves confirm the significant differences in the effect of the three measures on flood risk.

- In the case of the rehabilitated old canal, the modified risk curve diverges significantly from the curve corresponding to the base scenario for return periods in-between 100 and 200 years approximately. For lower flood frequencies, the modified curve remains lower than in the base scenario. This is consistent with the hydraulic effect described above.
- For the recalibrated floodplain, the reduction in potential damage is found to monotonously increase as the flood frequency decreases.

- In contrast, in the case of heightening of existing walls, the potential damage is reduced for flood events with a relatively high flood frequency, whereas the modified risk curve crosses the curve corresponding to the base scenario for a return period close to 100 years. For more extreme events, the modified curve even exceeds the present one, consistently with the hydraulic effect detailed above.

Those details remain undisclosed if the differences between the risk curves are aggregated into a single number, such as the annual avoided risk (Figure 2).

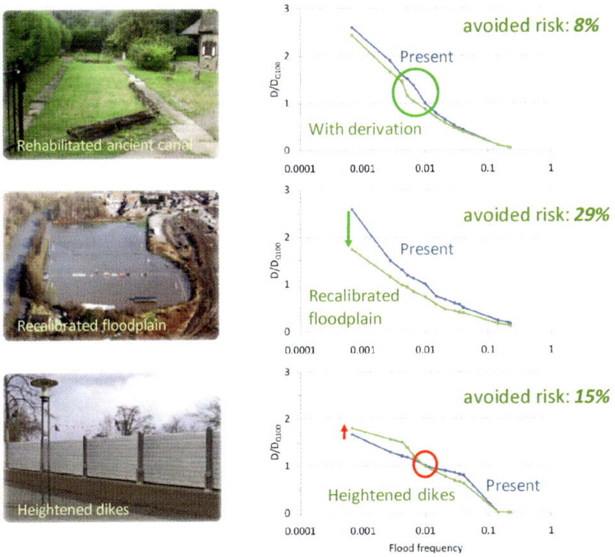

Figure 2: The risk curves reveal that the three considered protection measures have a very different influence on flood risk.

5 Conclusion

The section presents a flood risk modelling procedure developed at the University of Liege. It enables both meso-scale and micro-scale analyses. The focus has been set here on the micro-scale approach. It relies on detailed two-dimensional hydraulic modelling based on Digital Surface Models representing each building individually (resolution: 2m × 2m). Complementary landuse maps, statistical database and other geographical data have been combined to identify each affected asset (building or facility) and characterize their nature. Next, state-of-the-art flood loss models have been used to estimate direct potential damage to housing, which is known to be the sector having mostly contributed to the overall losses during recent flood events in the area of the considered case study. Finally, a recent methodology to quantify the social vulnerability of people and the adaptive capacity of communities has been used to estimate the intangible (psycho-social) effect of floods.

As demonstrated through a case study, such risk-oriented analysis discloses findings which would not arise from a more standard hydraulic study such as based on a design flood or return period. In particular, a single risk number has been shown not to be enough to communicate the concept of risk. The whole risk curve (flood frequency versus potential damages) needs to be analyzed to reveal in which range of flood events a protection measure is beneficial or not.

The presented micro-scale flood risk model is readily available to conduct flood risk analysis, feed cost-benefit analyses, as well as evaluate protection measures and hydrological effects of climate change scenarios. Scenarios of landuse change and environmental impacts are being incorporated in the risk model.

References

[1] Apel, H., Aronica, G., Kreibich, H., and Thieken, A. (2009). "Flood risk analyses - how detailed do we need to be?" *Nat. Hazards*, 49(1), 79-98.

[2] Coninx, I., and Bachus, K. (2007). "Integrating social vulnerability to floods in a climate change context." *Proc. Int. Conf. on adaptive and integrated water management, coping with complexity and uncertainty*, Basel, Switzerland.

[3] Coninx, I., Dewals, B., Detrembleur, S., Erpicum, S., and Pirotton, M. (2011). "Social flood risks quantification for use in spatial decision support systems: measuring the immeasurable?" In preparation. Information available from the authors on simple request.

[4] Dewals, B. J., Erpicum, S., Detrembleur, S., Archambeau, P., and Pirotton, M. (2011). "Failure of dams arranged in series or in complex." *Natural Hazards*, 56(3), 917-939.

[5] Ernst, J., Dewals, B. J., Detrembleur, S., Archambeau, P., Erpicum, S., and Pirotton, M. (2010). "Micro-scale flood risk analysis based on detailed 2D hydraulic modelling and high resolution geographic data." *Nat. Hazards*, 55(2), 181-209.

[6] Erpicum, S., Dewals, B. J., Archambeau, P., Detrembleur, S., and Pirotton, M. (2010a). "Detailed inundation modelling using high resolution DEMs." *Engineering Applications of Computational Fluid Mechanics*, 4(2), 196-208.

[7] Erpicum, S., Dewals, B. J., Archambeau, P., and Pirotton, M. (2010b). "Dam-break flow computation based on an efficient flux-vector splitting." *Journal of Computational and Applied Mathematics*, 234(7), 2143-2151.

[8] EU. (2007). "Directive 2007/60/EC of the European Parliament and of the Council of 23 October 2007 on the assessment and management of flood risks." Commission of the European communities (EU), Brussels, Belgium.

[9] Kaplan, S., and Garrick, B. J. (1981). "On The Quantitative Definition of Risk." *Risk Analysis*, 1(1), 11-27.

[10] Kreibich, H., Thieken, A. H., Petrow, T., Müller, M., and Merz, B. (2005). "Flood loss reduction of private households due to building

precautionary measures ? lessons learned from the Elbe flood in August 2002." *Nat. Hazards Earth Syst. Sci.*, 5(1), 117-126.

[11] Roger, S., Dewals, B. J., Erpicum, S., Schwanenberg, D., Schüttrumpf, H., Köngeter, J., and Pirotton, M. (2009). "Experimental und numerical investigations of dike-break induced flows." *J. Hydraul. Res.*, 47(3), 349-359.

[12] Tapsell, S. M., Penning-Rowsell, E. C., Tunstall, S. M., and Wilson, T. L. (2002). "Vulnerability to flooding: health and social dimensions." *Phil. Trans. R. Soc. Lond.*, A(360), 1511-1525.

[13] Thieken, A. H., Müller, M., Kreibich, H., and Merz, B. (2005). "Flood damage and influencing factors: New insights from the August 2002 flood in Germany." *Water Resour. Res.*, 41(W12430).

[14] WG F. (2011). "Resource document on flood risk management, economics and decision making support." Working Group F of the Common Implementation Strategy for the Water Framework Directive.

Real-time environmental information for the public and public safety organisations in case of flood events

S. Gottwald[1] & P. Hasenfuss[2]
[1]*Capgemini, Germany*
[2]*PSI Transcom GmbH, Germany*

Abstract

Recent flood events across Europe have resulted in the development of new applications for high water situations. Real-time environmental monitoring systems prove increasingly valuable while meteorological conditions are changing worldwide. The 2005 and 2006 floods in southern and eastern Germany showed the importance of fast reaction in civil protection. The period needed for alerting the public can be significantly cut by real-time systems. These have to be embedded in a communication infrastructure with clearly defined responsibilities, short information paths, and the acceptance of contemporary technologies to help mitigate the disastrous impacts of floods.
Keywords: flood prevention, civil protection, forecasting, alert system, monitoring system, real-time information delivery, communication infrastructure.

1 Introduction

High water situations are getting more and more threatening since they occur more and more frequently. Therefore people today are interested in water levels near their home. But it has to be pointed out, that the public view is much more focussed on large rivers and places which are more or less famous or well-known. However, high water is much more dangerous along small river basins as the period of time between an intense rain and the appearance of a flash flood amounts to just a few hours. So public services have not only to take care about physical flood protection but also the awareness of people near such small rivers, which never thought about rising water levels ten times higher than they were usually.

When reviewing flood events over the last years then a lot of research has been triggered and new conclusions have been implemented (latest discussion published e.g. in [1–3]), such as,

- research on impacts of global climate change regarding precipitation and snow and glacier smelt;
- research on and development of regional precipitation forecasts;
- investigation and implementation of physical flood protection like building or renewing dams, dikes, sluices or renaturation of river beds;
- build-up or redefinition of information processes and infrastructure;
- acceleration and automation of information processes;
- building public awareness.

With this input from scientific research results and experiences from public safety organisations (PSO) PSI started to build up a new information technology to shorten the period of time between measurement of rising water levels, forecast and actions of PSO and information of the public. To explain, that this is in case of Germany not just a technological problem, but also a political one, we have to have a quick look on the key players in flood protection in Germany.

1.1 Flood protection authorities in Germany

The nation of Germany is a federally organized country, split up in 16 federal states (German: Bundesländer). Disaster protection and all kinds of environmental issues, e.g. flood monitoring are within the responsibility of these federal states. But all topics about disaster protection belong to the Ministry of Interior, while the environmental issues belong to the Ministry of Environment. As fortunately not every accident became a large scale disaster the federal states delegated responsibilities in disaster management to the administrative districts (German: Landkreise). On this political level fire workers, rescue services, police and majors deal with disaster management. Each of the administrative districts and federal states, too, focuses, of course, on its own region first.

But nature does not think in terms of political regions, e.g. rivers follow their natural river beds. The Rhine River, one of the longest and most important rivers in Europe, flows through six European countries, five German federal states and lots of German administrative districts.

In consequence of financial and political matters each of the water units of the federal states develops its own flood level monitoring system and operates its own "flood centre". This leads to an open competition between civil authorities in developing the "best" system. Excellent systems are developed by many companies, but sometimes with a different focus. Most of them just provide the information of water levels rising or falling. If any head of a civil protection service want to get this information, he has to get it by himself, via internet, phone or data-transmission.

An answer to that sometimes for non-insiders inscrutable responsibilities is a "flood alerting and information order", as it is implemented in several federal states. It defines the process, how to get information from a "flood information centre" to the public and PSO.

1.2 Flood protection as a matter of time

In case of a flood event a large number of authorities responsible for civil protection have to be alerted quickly. The question of "Who is going to be informed when and by which communication channel by whom" is not only a highly political question but also a very difficult one to answer.

In most of the German municipalities home-grown communication and information technology already exists. At the information chain from gauges measurement technology to phone and fax availability there is a huge variety of technology development from the last 30 years. As a consequence it might happen that there is a perfect structured federal "flood alerting and information order", a modern type of gauge and a working "flood information centre", which cares about the measurement. But in case of a flood warning on Friday evening the information about an upcoming flood is located in front of a fax in the municipal office and nobody recognises until Monday morning.

Building up an information system that monitors water levels in different catchment areas and automatically alerts responsible authorities requires a communication infrastructure that goes far beyond the normal standards in information technology (IT). PSI build up a system like this with "PSI*econtrol*". Also, the realization with IT has its limits in the event of a real catastrophe.

The quality of real-time system integration is shown by operations handled every day in different communication infrastructures. Where flooding occurs the time between alerting and the actually rising flood is decisive:

- How much time remains for additional flood prevention measures?
- How long is the information chain between recognition of a threatening flood and the last persons affected by it, the citizens?

The period of time between recognition and actual flood depends on the geographic situation, too: in highlands floodwaters rise much faster than in lowlands. In case of a threatening flood time is the most important factor: the period of time between the serious warning that a flood will come and the arrival of the water is the time that can be used for additional measures against the flood.

2 Information process

The general information process with PSI*econtrol* in flood protection is rather straightforward:

- data collection (precipitation, water level, discharge, etc.);
- real-time monitoring of the data;
- alerting (if predefined alarm-levels are exceeded);
- action management by the flood protection units.

There are quite a number of participants involved with various responsibilities such as to supply measurements and prognoses or make decisions and carry out

action plans and orders. To give the key player a comprehensive support at their job PSI developed an innovative control centre system called PSI*econtrol*. PSI*econtrol* is field-tested for supra-regional flood management. Through use of sophisticated modern technology it enables the integration and monitoring of diverse environmental data in real time and when necessary it will initiate detailed alarms.
The key features of the system are:

- fail-safe and failure tolerant system;
- real-time application;
- simple operation;
- fully automated measurement and alarming;
- visual monitoring of message-handling, alarming and archiving.

2.1 Data collection

At the start of this information chain weather services play an important part as they forecast precipitation, temperature and sunshine duration. The better the location and the rain intensity are predicted the earlier and more detailed water levels can be computed. In highlands local precipitation forecasts are a real challenge and their impacts can be immense – strong rain on the one or the other side of a hill causes floods in completely different valleys and river catchments. Therefore, precipitation measurements are necessary and decisive in such sensitive areas.

This leads to the next complex task – data collection. Water levels and discharge measurements have to be collected automatically at short time intervals, e.g. every 15 minutes and runoff data have to be computed. As rivers and river systems usually pass different administrative borders a quick and comprehensive cross border data exchange has to be established. In some areas additional human observers are integrated into the information process. These observers look after a water pole and take additional measurements to ensure the quality of all electronic measurements. During past flood events in some cases human observers had to take over completely and provide the measurements since whole measuring stations were flushed away. They can also provide additional observations on water discharge like icing or flotsam washed ashore.

2.2 Monitoring

All measurements and observations from all stations have to be checked for measuring or data transmission errors so that only feasible data are stored and used for further processing and actions. A real-time monitoring system is used to check for limiting values, to trigger automatic alarm events and to visualize all necessary data. Based on actual precipitation and runoff data and time series of past days or weeks prognoses of future water levels can be computed.

In case of a threatening or genuine disaster it is important that the hydrologist on duty is provided with general overviews of complete river catchments as well as detailed views of the development of a flood in a specified area. He needs this information to enable him to compile a position report integrating the meteorological and hydrological situation and further developments.

2.3 Alerting and information delivery

As far as the information and communication process is concerned all data and information are available at this point. Now the decisive steps have to be taken: the information must be delivered to all persons concerned. As pointed out in the introduction an information and communication model has to be established stating which information has to be delivered to whom.

Such complex information and communication model can only be implemented when using a powerful and flexible software system which allows an adequate mapping of the defined processes, e.g. comprehensive profile management, individual availability, defining Alarm Channels by Time of Day and Means of Alerting.

And complete automation of the information processes is the key to significantly speeding up the process of warning and alerting citizens and it is the basis of an early warning system. With PSI*econtrol* information is delivered through different communication channels. Usually it is sufficient to inform the public on a web page or teletext. Depending on gravity, urgency, region, incidence, time of the year or time of the day it is possible to compile the information and deliver directly via the existing information channels such as fax, email, SMS, radio communication of paging systems.

2.4 Managing the flood: emergency plans

The last step of the information process presented here consists of the assistance of authorities responsible to manage the flood. In each area potentially affected by a flood action and emergency plans are compiled. These plans usually refer to alarm or water levels of a nearby water level gauge.

Having a communication infrastructure as mentioned above it is only a small technical step to store and maintain these action plans within a software system which automatically alerts recipient groups to execute predefined actions, e.g. road blocks or evacuation of vulnerable places at a certain water level. Again, actions can bc monitored and controlled and manual feedback can be integrated into an information process.

With the use of a real-time monitoring and alerting system the period of time can be fully utilized. At the time the flood forecast recognizes an upcoming flood situation on account of water levels and precipitation the information process is set off automatically without any human interaction regardless of the time of day. All relevant persons are informed via their preferred communication channels. Thus the remaining time until the water arrives can be optimized

3 Example Free State of Saxony

3.1 The situation in 2002

The flood event in the German Free State of Saxony in summer 2002 did not only leave much damage behind in cities and the countryside in Saxony, it also destroyed the existing infrastructure used for flood warning. The flood warning system was literally flooded away. Up to that time Saxony had 3 flood warning centres that were responsible for different river basins. The need for a new technical infrastructure was a chance to renew the organisational structure as well and to introduce forward-looking technical approaches.

Figure 1: Dresden/Saxony in 2006.

3.2 Renewing the state's flood water centre

The renewal of the destroyed flood water infrastructure was accompanied by renewing the whole organization and technique. Initially, the three flood water centres were united and the information infrastructure was renewed.

Having one centre for the whole federal state of Saxony also implied that all person-to-person relationships of the old system had to be replaced by machine-man communication. The centre simply does not know any more the persons responsible for taking action in the event of a (threatening) flood.

3.3 A technical solution

PSI implemented the technical solution "PSI*econtrol*" in Saxony. [4] The system is located in the state's flood control centre (Landeshochwasserzentrum). All rivers in Saxony are monitored in real time, predictions on water level developments are made with the help of weather forecasts by the German Weather Service (Deutscher Wetterdienst).

The System consists of three components: a control unit, an information management system and a communication platform.

- The control unit polls all available control points automatically. Data can be exchanged via voice, web or WAP with enablers on site.
- The information management system is the core of the whole system where all data and documents are stored in redundant and fail safe systems. Analyses, plausibility checks and monitoring of data from all control points are automatically carried out.
- The communication platform is responsible for informing, and when appropriate, initiating an alarm to the appropriate person(s) dependent upon the detected values.

Based on the definitions and experiences made in the German Free State of Saxony the following model is feasible, useful and approved. There are three main types of information or documents:

- management or position reports for each river catchment which covers the meteorological and hydrological situation and include recommendations for actions;
- water level reports which consist of a form containing actual measurements, trends and alarm levels;
- urgent short messages which signal the first-time overrun of the alarm threshold and are replicated in case of transgression of further alarm levels.

The communication model in Saxony is based on the following principles:

- all local partners with responsibilities like mayors, fire or water brigades, police and other safety or rescue services receive directly the above mentioned documents where relevant to their field of responsibility;
- the receipt of an urgent short message has to be confirmed manually so it can be assumed that the content has really been read;
- in case an urgent short message is not confirmed within a defined time span escalation reports are generated and delivered to the supervisory authority;
- all information can be delivered via different and redundant communication channels like fax, email, SMS and voice;
- the public is informed by automatically updated internet and teletext messages.

Manual intervention is also possible if, for example, the operator can anticipate a hazard. Extensive configuration parameters enable very flexible choices as to which information should be conveyed when and to whom and whether it should be passed via fax, email, SMS or voice.

The last major flood event managed by the system occurred in March and April 2006 in Saxony. It was caused by snowmelt and additional precipitation. Within a few hours the water levels of many usually very small rivers rose and flooded streets, villages and landscapes. In this critical situation PSI*econtrol* managed to issue up to 120 parallel alarms and to deliver up to 8.000 documents within 24 hours.

4 Conclusions for the future

IT solutions are really only a question of technology and can easily be found. A more complicated problem is the definition of communication channels and alert mechanisms. The introduction of systems like PSI*econtrol* seems to be more complex on the organisational rather than on the technological side.

PSI*econtrol* itself is rather independent of these questions: due to its adaptability and upgrading functions it can be configured rather easily. Integration of any other technical water level metering equipment with communication protocols is possible. These include alarm channels such as paging devices or even news tickers on TV programs and internet. GIS functionality will be employed for monitoring the alerts as well as for flood maps. PSI develops feasible extensions jointly with local authorities and customers.

The aggregation of various data may be subjected to real-time evaluation for risk analyses and be used for modelling escalation scenarios by an integrated solution. Data-input from air supported dike monitoring enables a real-time dike breach simulation as well as the checking of the water quality. At the same time it provides for harmonising GIS and alarm systems. In future, this information will soundly support necessary decisions.

PSI Environmental Disaster Protection Unit started a project to design a software system which will support hydrologists in flood warning centres as well as local civil protection authorities. It would thus be possible to manage water levels along the river (monitoring, forecasting, alerting) and to support decision makers like mayors in taking the right measures and monitoring progress and success.

References

[1] Morell, M. et al (eds). BALWOIS 2006 – Abstracts. *Conference on water observation and information system for decision support*. May 23-26, 2006, Ohrid, Republic of Macedonia. Hydrometeorological Service of Republic of Macedonia, 2006.

[2] Steckeweh, C. (eds). 3rd International Trade Fair and Congress for Flood Prevention, Consequences of Climate Change and Disaster Management. (Plenum II). *Flood Protection*. Hamburger Messe und Congress GmbH. Hamburg and Berlin. pp. 28–39, 2006.

[3] Steckeweh, C. (eds). 3rd International Trade Fair and Congress for Flood Prevention, Consequences of Climate Change and Disaster Management. (Forum 6). *Flood Protection in Municipal Practice*. Hamburger Messe und Congress GmbH. Hamburg and Berlin. pp. 96–103, 2006.

[4] Gottwald, S., High Water Monitoring and Warning via PSIecontrol. *InterCarto-InterGIS 12 – Proceedings. International Conference on Geoinformation for Sustainable Development. August 28-30, 2006 Berlin, Germany.* Eds Kremers, H. & Tikunov, V. Deutsche Gesellschaft für Kartographie e.V. Berlin, 2006.

Alternatives in flood protection: the effect of decentralized measures in the Upper Flöha watershed (Southeastern Germany)

Ch. Reinhardt, J. Bölscher, R. Imjela, M. Ramelow, R. Wenzel & A. Schulte
Institute of Geographical Sciences, Freie Universität Berlin, Germany

Abstract

In the German-Czech border regions throughout the Ore Mountains storm runoff frequently causes severe damage in headwater areas and lower reaches. The fact that settlements along smaller tributaries and towns near the receiving water are affected simultaneously requires a flood control concept with measures applied throughout the entire drainage area. Modern concepts of flood protection offer a large number of potential natural measures such as a number of small, decentralized retarding basins, river revitalization, afforestation of flood plains or land use changes. The study presented here is part of the Interreg-IIIA research project "DINGHO". Its aim is to show how decentralized measures can contribute to improving flood protection in the Upper Flöha catchment. The advantages and limits of such natural flood protection measures will be illustrated.

The study area extends between the large Rauschenbach reservoir and the town of Olbernhau (Saxony), including two main tributaries – the Natzschung and Schweinitz rivers. In a first step the local potentials were investigated on the basis of GIS analyses and field surveys. The second step included the assessment of the efficiency of the decentralized measures in different scenarios using the rainfall-runoff models NASIM and WaSiM-ETH.

Especially in higher regions with gentle slopes (above ~700 m a.s.l.), wide areas along the river channels can be used for water retention in small, decentralized retarding basins. In addition, some areas can be reforested, since they have recently recovered from the SO_2 pollution which affected the area until the late 1980s. By contrast, potentials for river channel renaturation are strongly limited.

The modeling results clearly indicate that small retarding basins would reduce peak discharges with a 100-year return period by up to 48% at some tributaries and 10% downstream in Olbernhau. River restoration measures do not show any significant effect. Reforestation of forest decline areas can reduce peak discharge by about 16%.

Hence an effective improvement of flood protection in the Upper Floeha catchment can be achieved through the application of decentralized measures. The reduction of flood peaks includes the Floeha river itself, tributaries and headwater areas.

This article is a reprint of the original version. However some modifications were included as far as the simulation of land use changes is concerned. The model used for this part was extended and improved, so that the text includes updates of the results.

Keywords: flood protection, decentralized measures, retarding basins, renaturation, reforestation of floodplains, land use changes, rainfall-runoff modelling.

1 Introduction

As shown by the August 2002 extreme flood and numerous local events described by Büttner *et al.* [1], losses arising from flooding caused by storm runoff frequently occur throughout the low mountain ranges of Southeastern Germany (Free State of Saxony) and especially in the Ore Mountains in the German-Czech border regions. Further deterioration has also to be considered because an increase in the frequency of storm events in the months July and August as a result of global climatic changes was predicted for Saxony (SMUL [2]). The strategy for adaptation and mitigation of flood risks in the Free State is based on a combination of precaution (early warning systems, education, etc.) and spatial planning (e.g. identification of flood generating areas, designation of inundation areas), widespread retention in catchment areas (e.g. funding of agricultural techniques that facilitate infiltration), as well as technical flood protection. To improve technical flood control, 47 individual concepts with a large number of proposals for defence construction projects were developed for first order rivers (Socher *et al.* [3]). By contrast, further research has to be carried out for second order rivers – where, according to the Water Act of Saxony, municipalities are responsible for water management and flood control – because protection plans are lacking in many cases.

Against this background, potentials for an improvement of flood protection in the Upper Flöha watershed were investigated in the framework of the "DINGHO" research project, which received funding from the Interreg-IIIA program of the European Union. The aims of the project are to reveal the contribution of decentralized and natural measures for flood control in municipalities along second order rivers (headwater areas) as well as downstream at the river Flöha. The studies focused on two different fields: 1.) measures along rivers and creeks (small retarding basins, river renaturation, afforestation of floodplains) and 2.) widespread land use and utilization changes. This article contains a summary of the main research results. A more in-depth description is given e.g. by Reinhardt [5].

1.1 The principle of decentralized flood protection

Since the early 1990s numerous studies on the effect of decentralized measures in different river basins have been published in the German literature (e.g. Wegner [4], Assmann *et al.* [6], Marenbach and Koehler [7], Röttcher and Tönsmann [8], Röttcher [9], as well as summaries edited by Röttcher *et al.* [10] or published by DWA [11]). The concept is based on the idea that in mesoscale river basins (≤1000 km³) a system of numerous small and distributed flood control measures can be established throughout the entire drainage area (fig. 1) instead of large defence constructions downstream (e.g. a large retarding basin). By distributing the measures an improvement of flood protection can be achieved that includes settlements in headwater areas as well as at the receiving water.

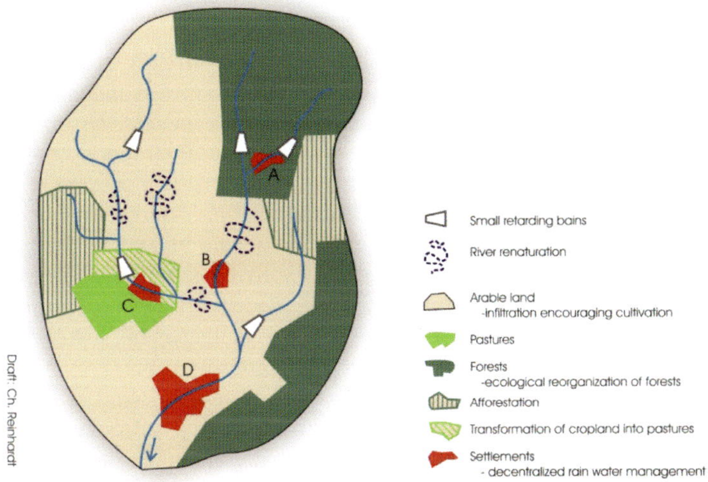

Figure 1: The principle of decentralized flood protection, illustrated for selected measures in a small watershed. A-C: municipalities located upstream and in headwater areas, D: municipality at the lower course of the studied river.

The large number of published decentralized measures can be classified into three different groups:

1.) Measures at and around rivers: e.g. systems of small retarding basins with a maximum storage capacity of 50,000–100,000 m³ each, retention behind obstacles (e.g. cross-valley road embankments), retention in existing ponds, river renaturation, reforestation of floodplains

2.) Watershed-wide retention e.g. afforestation, ecological reorganization of forests, conversion of arable land into pastures, infiltration promoting cultivation of croplands

3.) Settlements: e.g. decentralized rain water management, renaturation of paved urban areas

Depending on the type and number of measures as well as the local characteristics of the drainage area under investigation, the predicted reduction of peak discharges for periodic flooding (return period of 100 years) ranges from only 1 to almost 50 percent (Bauna river, Röttcher and Tönsmann [8]). The previously published data show clearly, that decentralized measures can be very efficient elements in flood control concepts. However, the large range of calculated peak reductions demonstrates that existing results cannot be assigned completely to other drainage areas. Different drainage structures lead to an individual combination and overlap of flood waves, so that separate investigations for any other river basin are required during the planning process.

Another important fact that has to be considered in the framework of flood control concepts is the potential impact of the planned measures on landscape and ecosystems. Whereas large technical constructions strongly influence natural systems along rivers, the impact of decentralized measures is much lower. The water level in retarding basins is restricted to a few metres, so impoundage remains short and negative influences on the natural scenery are limited. Moreover, the renaturation of river systems and associated measures lead to an activation and enforcement of natural retention in floodplains, ecological demands being satisfied at the same time. Land use and utilization changes may encourage infiltration and cause widespread improvements of the natural water balance. Owing to these synergistic effects, decentralized measures can be understood as part of an integrated watershed management, conforming to requirements of flood protection as well as nature conservancy.

In the international literature similar approaches can be found in the context of watershed management and rainwater harvesting (cf. Brooks et al. [12], Lancaster [13], Westervelt [14]). However, in many cases the basic ideas, aims and the application of the concepts are not completely identical.

2 Study area

The study area is located 50 km southwest of Dresden in the central Ore Mountains and includes the Upper Flöha river between the large Rauschenbach reservoir and the city of Olbernhau (fig. 2). With its two main tributaries, the Natzschung and Schweinitz rivers, the watershed extends over an area of 228 km². Because time series of the gauge in Olbernhau for the calibration of the rainfall-runoff models have been available solely since 2005, the modeling area had to be enlarged downstream towards the Pockau 1 gauge (315 km²). The low mountain geomorphology is characterized by two main landscape types. Headwater areas, which are located in higher regions, are developed plateau-like

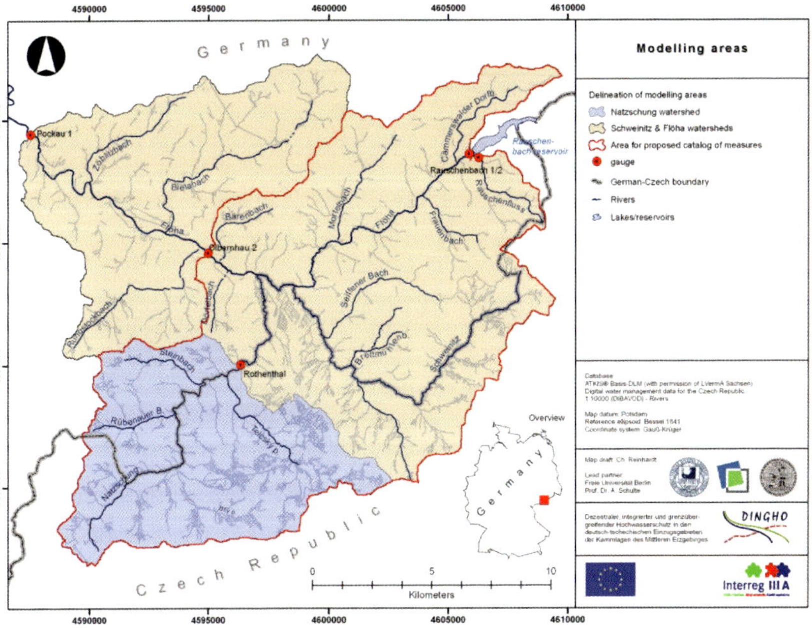

Figure 2: The Upper Flöha basin between the Rauschenbach reservoir and the Pockau 1 gauge including the modeling areas Natzschung and Schweinitz/Flöha. The area for the proposed measures (red outline) is located upstream of the town of Olbernhau (Olbernhau 2 gauge).

with wide valleys and gentle slopes (remnants of peneplains). By contrast, lower reaches are strongly dissected by steep V-shaped canyons, incised during the Quaternary.

Approximately two-thirds of the area under investigation is covered by forests, including coniferous (28% of the total catchment area), broad-leaved (6%) and mixed forest (8%). Another 24% are characterized by a transitional woodland-shrub-like vegetation. In the Ore Mountains this vegetation type is a result of the SO_2-pollution-related forest decline, which affected higher mountain reaches until the late 1980s and early 1990s. Because these regions are recovering gradually they are subject to reforestation with spruce and secondarily to natural succession with birch and rowan. The remaining third of the total area is used for agriculture (cropland, pastures), whereas settlements cover 3 percent only.

3 Rainfall-runoff modeling

The analysis of the effects of decentralized measures in the study area is based on the application of distributed rainfall-runoff models. Depending on the type of measures under investigation, two different model systems were used. Measures along rivers (retarding basins, renaturation and afforestation of floodplains) were investigated using the model system NASIM, designed by Hydrotec (Aachen, Germany). As shown in figure 2, the total modeling area was subdivided into two independent components – the Natzschung watershed model (Bölscher [15]) and the Schweinitz/Flöha model (Reinhardt [5]). To assess the total effect of all measures in Olbernhau, the models were connected at the gauging station of Rothenthal (fig. 1). By contrast, the influence of land use and utilization changes was investigated for the entire study area using the model system WaSiM-ETH.

To calibrate the hydrological models, time series of discharge for the period 2002-2006 from Rothenthal/Natzschung, Pockau 1/Flöha and Olbernhau 2/Flöha gauges were used. On the basis of the calibrated models and statistical rainfall amounts, a design flood event with a return period of 100 years was simulated in a second phase, which serves as a reference for assessing the efficiency of the planned measures. During the third working phase, different scenarios for decentralized measures were implemented in the watershed models. The process of scenario development incorporated various results from field studies, e.g. potential location and storage capacity of the retarding basins, which were determined by geodetic surveys, or river sections for renaturation which are based on detailed water structure surveys.

4 Results and discussion

4.1 Potentials for decentralized measures

Depending on the class of measures, the potentials in the study area differ strongly. In the Natzschung watershed (fig. 2) 19 locations for small retarding basins were identified, including 3 artificial ponds which can be used for flood protection by lowering the existing water level. Additionally 13 sites throughout the model area of Schweinitz/Upper Flöha were considered, resulting in a total of 32 basins with a cumulated storage capacity of nearly 800,000 m³. The storage level in the retarding basins is limited to 4–5 m. To allow the constructions to be fully integrated into the natural scenery, earth dams with a cross section gradient of 1:3 were assumed. Discharge is controlled by unregulated throttles with diameters of 300–1,800 mm, which can be constructed as round concrete pipes.

According to the water structure surveys and the mapped state of the rivers, only very few short sections can be considered for renaturation. By contrast, potentials for the afforestation of floodplains can be found along almost all water courses, so these measures were considered along all river sections, which are located outside settlements.

As far as land use changes are concerned, these measures can be realized in the long term only, so in this article we focus on scenarios, which seem to be realizable. For the area under investigation such potentials can be found mainly in the afforestation of regions that were affected by heavy forest decline until the late 1980s. To understand the general hydrological effects, additional scenarios for a complete forest cover of the entire study area and the conversion of croplands into pastures will be presented.

4.2 Hydrological effects

The proposed scenarios show a strong variability in their hydrological effects, depending mainly on the type of measures and the reference location. For municipalities in headwater areas along second order rivers, a high reduction of peak discharges of the reference flood event (100-year recurrence interval) by the construction of retarding basins can be registered, e.g. in Rübenau (45%), Načetin/Czech Republic (48%, Bölscher [15]) as well as in Seiffen (28%, Reinhardt [5]). For the city of Olbernhau, in the lower reaches of the study area, the model reveals a possible reduction of 10% (fig. 3) when all 32 retarding basins are included. Modeled separately the 19 retarding basins throughout the Natzschung watershed would lead to a reduction of 6% in Olbernhau, whereas the 13 sites throughout the modeling area of Schweinitz/Flöha may reduce peak discharges by 4%.

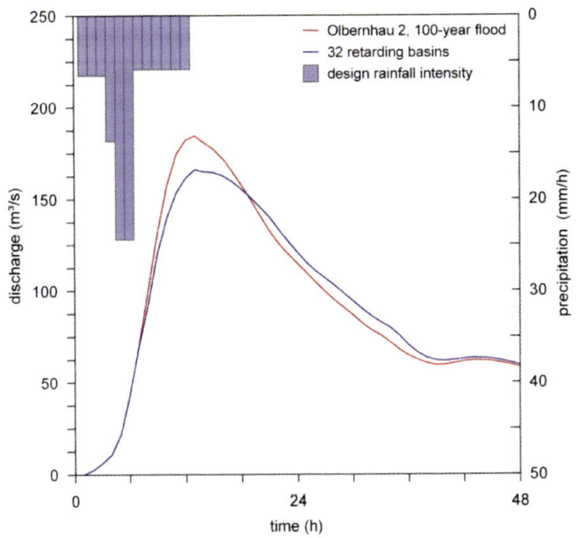

Figure 3: Hydrographs showing the effects of the proposed decentralized retarding basins for the city of Olbernhau at the river Flöha.

In contrast to the clear effects of the retarding basins, the considered river renaturations do not result in a significant reduction of flood peak discharges – neither for Olbernhau nor municipalities in headwater areas. However, low efficiency is not a general problem of renaturation, but rather caused by the limited number of river sections in the study area, that are suitable for such measures. Distinct effects can be expected from the afforestation of floodplains, which result in increased roughness coefficients along the rivers. Under consideration of afforested floodplains in the watersheds of Schweinitz and Flöha, a peak reduction of 4.3% and a lag of about 2 hours can be achieved for

Olbernhau. The main retention effects occur along the Flöha river (3.5%), where the bed slope is low and the floodplain wide, whereas afforestation along tributaries leads to a further decrease caused by a slightly stronger separation of flood waves in the model. When the Natzschung is included, no further increase of the effect in Olbernhau can be noticed, although Bölscher [15] observed a significant reduction at the Rothenthal gauging station in his model of the Natzschung watershed.

Because significant hydrological effects for retarding basins as well as afforestation of floodplains were registered, the combination of both measures was also examined. According to the modeling results in Olbernhau, a peak discharge reduction of 6.6% can be achieved when combined measures in the Natzschung watershed are considered, whereas measures in the watersheds of Flöha and Schweinitz would lower flood peaks by 8.2%. Combined measures in both modeling areas would result in a reduction of 12.8% in Olbernhau (fig. 4).

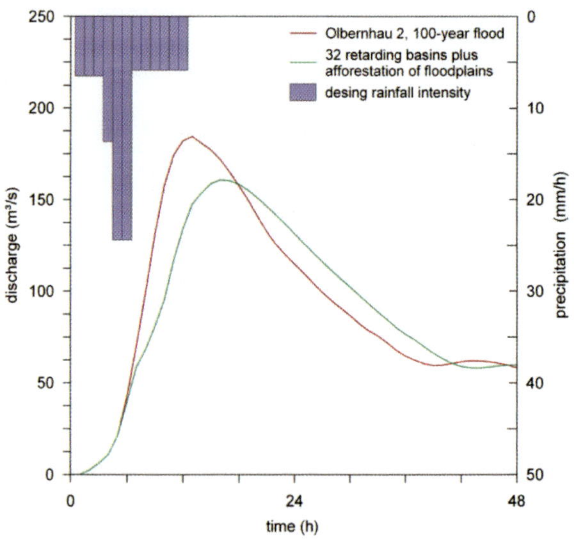

Figure 4: Hydrographs showing the effects of combined measures for the city of Olbernhau.

The modeled land use scenarios reveal reduction effects similar to those achievable by a combination of retarding basin and floodplain afforestation. If the recently declined forest areas (transitional woodland-shrub vegetation) were replaced by mixed or coniferous forests a flood peak reduction of 10% and 16.4% respectively occurred (fig. 5). By contrast, the transformation of croplands in the study area into pastures does not indicate any significant supralocal effects for summer flood events. However, when assessing the effect of land use changes it has to be considered that in regions with intensive

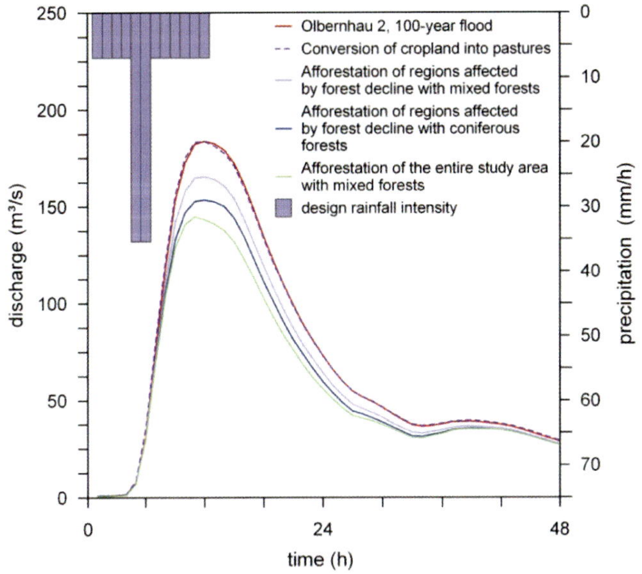

Figure 5: Hydrological effects of land use changes on flood peaks in Olbernhau for different model scenarios.

agriculture and forestry a long period is required for planning and the transformation process. Against this background the systematic afforestation of forest decline areas and the development of mixed forests offer chances to improve flood protection as well as natural habitats without high additional investments.

5 Conclusions

The modeling results clearly indicate to what extent decentralized measures contribute to an improvement of flood protection in the area under investigation (tab. 1). Beneficiaries of small, natural retarding basins include settlements along the Flöha tributaries, where very high flood peak reductions of up to 48% were observed by the application of the rainfall-runoff models. Beyond tributaries and headwater areas the effects along the receiving water (Flöha) are lower, but still reach 10% in Olbernhau. In combination with a reinforcement of the retention along the river by afforestation of floodplains, peak discharges can be decreased by 12.8%. Similar reductions can be achieved by afforestation of declined forest areas, when the newly established tree population has developed into a closed forest system.

On the whole, we may conclude that decentralized measures offer good opportunities to bring together flood protection and ecological aspects. Apart from the recognized hydrological effects, expressed by the modeled flood peak reductions, the impact on the natural system is very low. On the contrary, various measures even lead to an improvement of the ecological state in mesoscale river basins. However, their applicability as flood protection elements strongly depends on the potentials and resources in the examined watershed, e.g. total storage capacity, number of river sections for renaturation, access to areas for land use changes, etc. If large potentials are available, an effective and natural flood protection can be established by decentralized measures.

Table 1: Summary of predicted flood peak reductions.

Measures	Flood peak reduction (%)
Small retarding basins	
Načetin	48
Rübenau	45
Seiffen	28
Obernhau	9.9
Afforestation of floodplains	
Olbernhau	4.2
Combination of retarding basins and afforestation of floodplains	
Olbernhau	12.8
Afforestation of regions affected by forest decline with mixed/coniferous forests	
Olbernhau	10.0/16.4
Afforestation of the entire study area with mixed forests	
Olbernhau	22.0
Conversion of cropland into pastures	
Olbernhau	0.1

References

[1] Büttner, U., Fügner, D. & Winkler, U., The flood on 5 and 6 July 1999 in the region of Marienberg in the Ore Mountains (Saxony). *Hydrologie & Wasserbewirtschaftung* **45(3)**, pp. 102-111, 2001. In German with an English summary.

[2] SMUL (Saxon State Ministry of the Environment and Agriculture), *Climate change in Saxony – state of research and perspectives*, Dresden, pp. 1-111, 2005. In German.

[3] Socher, M., Dornack, S. & Defér, E., Flood-protection concepts in the Free State of Saxony – an introduction. *Hydrologie & Wasserbewirtschaftung* **50 (6/2006)**, pp. 303-308, 2006. In German.

[4] Wegner, H., Decentralized flood control. *Wasser & Boden* **44 (1/1992)**, pp. 6-10, 1992. In German with an English summary.

[5] Reinhardt, C., Decentralized flood protection in the Upper Floeha watershed (Central Ore Mountains), PhD thesis, Department of Earth Sciences Freie Universität Berlin, Shaker Verlag, Aachen, pp. 1-192, 2010. In German with an English summary.

[6] Assmann, A., Gündra, H., Schukraft, G. & Schulte, A, Decentralized and integrated flood protection planning for the Upper Elsenz. *Wasser & Boden* **50(8)**, pp. 15-19, 1998. In German with an English summary.

[7] Marenbach, B. & Koehler, G., Local and supralocal effects of central and decentralized measures of flood protection. *Wasser & Boden* **55 (7/8)**, pp. 22-26, 2003. In German with an English summary.

[8] Röttcher, K. & Tönsmann, F., Potentials and limitations of a decentralised flood protection approach in the German Mittelgebirge: the Bauna River case study. *Wasser & Boden* **53(10)**, pp. 29-34, 2001. In German with an English summary.

[9] Röttcher, K., Decentralized and central water retention measures for flood protection – exemplified investigations at the Erpe River. *Wasser und Abfall* **5**, pp. 11-15, 2005. In German.

[10] Röttcher, K., Koehler, G., Kleeberg, H.-B. (eds), *Decentralized flood protection*, Forum für Hydrologie und Wasserbewirtschaftung **17.06**, pp. 1-182. In German.

[11] DWA (German Association for Water, Wastewater and Waste), *DWA-topics – Decentralized measures for flood reduction*, Hennef, pp. 1-110, 2006. In German.

[12] Brooks, K., Ffolliot, P.F., Gregersen, H.M., DeBano, L., *Hydrology and the management of watersheds,* Iowa State University Press, Ames, pp. 1-502, 1997.

[13] Lancaster, B., *Rainwater Harvesting for Drylands and Beyond, Volume 2: Water Harvesting Earthworks*, Rainsource Press, Tucson, pp. 1-417, 2008.

[14] Westervelt, J., *Simulation Modeling for Watershed Management*, Springer, New York, Berlin, pp. 1-190, 2001.

[15] Bölscher, J.: *The analysis of potentials for decentralized water retention in the Natzschung watershed (Central Ore Mountains) using the rainfall-runoff model NASIM*, PhD thesis, Department of Earth Sciences Freie Universität Berlin, in prep. In German with an English summary.

Psychological factors affecting flood coping strategies

C. B. Rose[1], K. Manktelow[2], C. A. Booth[1] & D. G. Proverbs[3]
[1]School of Technology, University of Wolverhampton, UK
[2]School of Applied Sciences, University of Wolverhampton, UK
[3]Faculty of Environment and Technology, University of West of England, UK

Abstract

There is an acknowledged need to improve the resilience of those at risk of flooding in areas of the UK. Studies of disaster preparedness worldwide indicate raising awareness of a hazard does not necessarily engender action. In the UK the majority of the at-risk population do not display adaptation behaviours until they have experienced one or more flood events, a finding not adequately explained by levels of information provision or financial pressures. An appreciation of the psychological underpinning of current behaviour patterns, including decision-making processes, can illuminate our understanding of the strategies employed. A review of relevant psychological theories affecting adaptation behaviours in a variety of hazard situations is presented. Findings reveal the influence of belief systems, such as locus of control and self-efficacy, on decision-making in risk environments: statistically significant correlations have been noted between the types of hazard-behaviours displayed and the scores obtained on the relevant belief system metrics. It is suggested psychometric measures might be employed as capability indicators and adoption of such techniques may contribute to improved resilience in the future.
Keywords: flood, resilience, psychology, preparedness, adaptation, capability.

1 Introduction

In the context of anticipated climate change [1], the UK government has enacted legislation in the form of the Climate Change Act [2]: increasingly worldwide attention is focussed upon assessing vulnerability to extreme events, improving upon recovery plans to deal with their aftermath and examining the ways in which adaptation to climate change might enhance resilience in the future. The overall picture is, however, a very complex one: the interactions between human society and the physical environment, whether natural or man-made, are open to interpretation from many different disciplines, and an accepted comprehensive model, integrating inputs from the physical sciences, engineering, socio-economic and organisational aspects, has yet to be formulated [3]. The issue of differential vulnerability illustrates this point, in that all the people who reside in a particular hazard zone are not necessarily equally vulnerable to event impacts [4], not only for reasons of socio-economic disadvantage, but also their ability to cope with post-event factors, such as psychological trauma. A full understanding of the ways in which people respond to hazard information cannot, therefore, be based solely upon a geographically bounded concept of what constitutes a 'community': the behaviour patterns and motivations of individuals making up that community must be considered [5] along with the ways in which behavioural changes can be elicited. To explore this topic more fully, we will now look at flood-related issues within the UK.

1.1 Awareness raising

Campaigns intended to promote protective behaviours via awareness-raising have been widely used, but may not always be as successful as the policy-making bodies would hope or expect: an annual campaign on flood-risk within England and Wales has been in place since 2001, but it has been found that only 60% of at-risk residents claimed to be aware they lived in a flood risk area [6]. Similarly, a free telephone flood warning system for at-risk properties has been available throughout this period, but at the time of the summer 2007 flood events, only 41% of eligible homes were registered for the Floodline Warnings Direct service [7]. Such behaviour patterns are not unique to the UK, nor are they peculiar to flood risk, as the literature on natural hazards worldwide will reveal [8–11]. The consistency of these findings (that awareness does not engender action) suggests that, in order to change human behaviour patterns, a more sophisticated approach may be required: this is discussed by Fischoff [12] in the context of risk communication strategies, in relation to a variety of hazards, over a twenty year period. If it is, therefore, accepted that appeals and advice predicated on rationality are insufficient in themselves, an examination of the psychological theories covering decision-making and behaviour change may offer a way forward in promoting resilient adaptation.

2 Psychological theories of decision-making

In the second half of the 20th century, psychologists initially viewed decision-making as being a predominantly rational process, involving logic, reason and consideration of all possible risks and benefits associated with a particular decision [13]. This 'rational actor paradigm' would, however, be an extremely time-consuming strategy, if employed afresh for each decision that human beings must undertake. It was, therefore, suggested that instead human beings commonly employ mental short-cuts, based upon experience, such as 'heuristics' to arrive at solutions swiftly [14]. Thus, when asked to estimate the probability of an event occurring, a person might employ the 'availability heuristic': they will make use of the associations most available, in that they can be brought to mind readily. Aircraft crashes, though in fact comparatively rare occurrences are easily brought to mind because they receive so much media exposure and are, therefore, thought to be more common than they really are. An alternative view [15] suggests human beings make use of 'fast and frugal' algorithms to make probabilistic judgements; for instance, memory is searched for cues to the problem and the first cue that is diagnostic will be adopted as the solution. Although such algorithms neither look up nor integrate all available information, empirical data [15] indicate that these strategies match, or outperform, all competitors in inferential speed and accuracy. As with heuristics, such mechanisms would provide a more time-efficient method of arriving at a conclusion, consistent with the cognitive limitations of human decision-makers.

3 Decision-making in risk environments

Although the above models can offer coherent explanations for many individual decision processes, and indeed some social choice situations, more complex mechanisms seem to pertain to human behaviour under conditions of uncertainty such as disaster preparedness. The 'psychometric paradigm', incorporating the heuristics argument, is a theoretical framework developed by Slovic and others over two decades [16, 17]. This assumes risk is fundamentally subjective, and defined by individuals in accord with the influence of psychological, social, institutional and cultural factors. In respect of natural hazards, the 'affect heuristic' was identified as an important factor: this short-cut makes use of the feelings associated with a given stimulus, for example liking the subject (positive affect) or disliking it (negative affect) [18]; this might colloquially be termed the 'gut reaction' to a situation.

The relationships between perceived risk, perceived benefit and risk acceptance are complex, but another consistent finding has been that, where perceived benefits are deemed to be high, the associated risks will be perceived as low and vice versa. This inverse correlation principle is harnessed by, notably, the advertising industry: product campaigns tend to emphasise the *benefits* of the goods or services offered, perhaps employing emotional content in order to do so; a health promotion campaign, however, is more likely to highlight the *risks* of a given behaviour, with the intention of deterrence or discouragement. A causal relationship here has been demonstrated [19], but careful consideration of behavioural theory is required, in order to identify the critical beliefs underlying the intentions of the target population if campaign materials are to be effective [20].

Another psychological theory of relevance here is prospect theory [21] which considers the importance of the 'framing' of problems, such as whether the result of a gamble will be a gain, or a loss, relative to the reference point of the current state. Expressing outcomes in terms of financial costs and benefits clearly resonates with the behaviour of residents in flood-risk areas, when they are considering investment in property-level resilience measures. Such decisions may also have timing implications: a dilemma can be seen to exist between short-term expenditure and long-term potential savings, and different groups of people will make different choices in the same situation. This issue may also be viewed in the context of motivational priorities [22]: it is in accordance with a rationality model that immediate and pressing needs would be addressed in preference to longer-term needs, especially if the latter are uncertain.

4 Social factors in decision-making

As social animals, our decision-making processes are influenced by the societies in which we live; there is the option to act as an individual, or to conform to a group norm. For example, a householder may wish to buy and install flood resilient products, in order to protect their own property; however, a dilemma can arise if s/he is aware this approach could attract opprobrium, from neighbours who believe advertising the existence of a flooding problem will affect the saleability of their own homes [23]. The ability to withstand social pressures, such as this, will vary between individuals, and indeed groups of people, bringing personality factors into the, already complex, problem. Other socio-cultural values, beliefs or superstitions exert effects on different sectors of the at-risk population [24–26].

The scope for misunderstandings in relation to flood risk has been well documented: for example, the misapprehension that dams actually built for hydro-electricity generation were flood-control structures [27]. A cultural example was noted in relation to the Easter 1998 flood event in Banbury, wherein some recent immigrants to the UK expressed surprise, as they had not expected to be flooded in a developed country [28]. Such belief systems, therefore, pose an additional contributory factor in decisions relating to flood risk adaptation. It has been noted that floodplain populations tend to ignore flood risk until either regulation, or repeated flood events, brings about behavioural changes [29]. A motivational mechanism for this phenomenon has been suggested: people without flood experience envisaged the consequences of a flood differently from people who had actually experienced severe flood losses [30]. In psychological terms, therefore, they underestimated the 'negative affect' associated with a flood: this has obvious implications for communication of warning campaigns across natural hazard scenarios, in that the potential for negative emotional consequences should be incorporated into educational materials, as well as information on mitigating tangible losses.

5 Measuring beliefs

Belief systems (a term that is not limited to religious beliefs in the present context) can have strong influences on decision-making processes; this applies not only in hazard adaptation contexts, but also in fields such as health, education and occupational psychology. Such beliefs may pose a barrier to action, even where awareness of hazard and knowledge of possible mitigation strategies both exist. Constructs offering explanations for belief-driven behaviours have been developed, as have techniques to measure these constructs; those used most widely in natural hazard studies will now be outlined.

5.1 Locus of control

This construct is derived from Rotter's Social Learning Theory [31]: where experience leads an individual to believe s/he is responsible for the outcomes of their actions, that person tends to develop an 'internal locus of control'. If forces external to the individual are perceived to be responsible for outcomes, however, the learning process is likely to result in the development of an 'external locus of control'. The Internal/External score (hereafter I/E) is typically measured via a forced choice expression of belief in a list of statements, such as:

a. Many of the unhappy things in people's lives are partly due to bad luck.
b. People's misfortunes result from the mistakes they make.

In an educational context, it has been found having an internal locus of control contributed to the educational attainment of US teenagers, operating via their expectations of the return on 'investment' represented by completing high school [32]. In studies of natural hazard behaviours, statistically significant correlations between behaviours and I/E scores were found in the context of earthquake preparedness [33]. It should be noted the locus of control orientation provides a measure of cross-situational beliefs, or *generalised* control beliefs, contrasting with the next concept to be examined

5.2 Self-efficacy

This concept refers to an individual's convictions (or confidence) about his or her abilities to mobilize the motivation, cognitive resources, and courses of action needed to successfully execute a *specific* task, as opposed to the more generalized locus of control scale discussed above. Self-efficacy, measured by means of a psychometric scale, has been utilised in human performance studies: employees who perceive themselves as highly efficacious will display a strong sense of commitment to their activities and, if these are well executed, produce successful outcomes; those who perceive their self-efficacy to be weaker will tend to lose confidence in their personal abilities and fail at set tasks [34].

5.3 Strength of belief

An additional characteristic of studies on adaptation to climate change is the inherent uncertainty around both the causes and, for some individuals and groups, doubt as to the existence of the problem itself. This is in direct contrast to models deriving from studies in the health sphere, where issues such as disease diagnosis are self-evidently real. The roles of strength of belief in climate change itself, and strength of belief in personal adaptive capacity, have been examined in the context of adaptations to forestry management in Sweden [35]. A significant positive association between the first characteristic and adaptation actions was identified, as was a significant association between lack of such belief and motivation for *not* adapting. They conclude the pre-requisites for positive adaptation were strong beliefs in the hazard itself, and the belief that, as an individual, a person has the power to do something about the hazard.

6 Next steps – is it possible to predict capability?

Many metrics of resilience have been identified (e.g. demographics; health and wellness; quality of life) but the majority of these apply at the community, rather than the individual level [36]. Such factors, therefore, can provide an overall indication of resilience in a geographically-bounded community; this is useful in monitoring the effects of a given intervention, but does not advance our understanding of how optimal interventions might be formulated and applied. Examination of factors affecting the individuals within a community may offer a more appropriate route for this; however, although suitable techniques exist for measuring such constructs, many of the conceptual models currently in existence are derived from the field of health studies. Such models may not be entirely appropriate when considering behaviours relating to climate change adaptations: for example, these lack the 'strength of belief' element discussed earlier. Utilising a suitable hazard-preparedness model can advance our understanding of the relevant behavioural factors at play in this specific context; such an approach offers the potential to predict community capability, based on the psychometric profile of the individuals within the group and some specific examples will now be considered.

In respect of disaster warning responses, consistent correlations have been noted between internal locus of control scores and behaviour; warnings need to be heard, believed, personalised and acted upon in order to be effective, and these behaviours are more likely to be found in internally, rather than externally, oriented people [37]. An investigation of disaster preparedness relating to hurricanes in the US, found the most important predictors of preparation behaviours were age, prior experience of a hurricane and locus of control score [38]. The self-efficacy construct has also been found to be of value in disaster resilience contexts: for example, it has been found that low self-efficacy scores equated to a feeling of being 'not competent to act' and, thus, the risks (in this case volcanic hazard effects) were perceived as insurmountable [8].

A German study of pro-environmental behaviour, using a model incorporating 'perceived behavioural control' (which includes a measure of self-efficacy) found this variable to be amongst the most powerful predictors of intention to act, explaining 52% of the intention construct [39]. Another German study makes use of a 'process model of private proactive adaptation to climate change' (MPPACC) [40]; this incorporates both self-efficacy and a factor termed 'fatalism', closely resembling an aspect of high externality, as key variables in the formation of adaptation intention within the individual. A subsequent modification to the model, specific to precautionary flood prevention in Germany, included the additional factor of 'threat experience appraisal', but this version omitted the impact of social discourse regarding climate change that had appeared in the earlier model [41]. It must be noted that cultural variations can impact upon the results of such studies; hence, findings from one nation's residents may not necessarily be replicated in other flood-risk populations [42]. An Australian model, originally formulated to account for Tsunami preparation behaviours, has been found to be applicable across other hazard environments [43], although the type of fluvial and coastal flooding typically found in the UK was not included. This model does, however, incorporate the concept of 'trust' as a mediating factor in the formation of intention, a variable not appearing in those discussed previously. It may, therefore, be argued that formulation of a model suitable for use with the flood-risk population of the UK is overdue.

7 Summary

This review has examined the psychological factors found to be of relevance in disaster preparedness, including the flood risk context. Drawing upon both decision-making theory and belief system constructs, the theories underpinning some current behavioural models have been described. The particular constraints operating in climate change related studies are found not to be reflected in those models derived from the health study field in the UK, whilst models predicated on climate change from Germany and Australia have not been tested within a UK flood-risk population. Further work to develop and test a model specific to flood preparedness behaviours in the UK is therefore required.

References

[1] UK Climate Impacts Programme, *UK Climate Impacts Programme,*. 2007.

[2] *Climate Change Act 2008.* 2008, HMSO: London.

[3] Mc Entire, D.A., C.G. Crocker, and E. Peters, *Addressing Vulnerability through an Integrated Approach.* International Journal of Disaster Resilience in the Built Environment, 2010. **1**(1): p. pp.50 - 64.

[4] Marchand, M., *Differential vulnerability in coastal communities: evidences and lessons learned from two deltas*, in *FRIAR (1st international conference on Flood Recovery, Innovation And Response) 2-3 July 2008* D. Proverbs, C. Brebbia, and E. Penning-Rowsell, Editors. 2008, WIT Press, Southampton: One Great George Street, London, UK. p. 283-293.

[5] Martin, I.M., H. Bender, and C. Raish, *What motivates individuals to protect themselves from risks: the case of wildland fires.* Risk Analysis, 2007. **27**(4): p. 887-900.

[6] Harries, T., *Householder responses to flood risk; the consequences of the search for ontological security.* 2007, Middlesex University: London. p. 255.

[7] Pitt, M., *The Pitt review, interim report.* 2007: H M Govt Cabinet Office, London.

[8] Paton, D., L.M. Smith, and D. Johnston, *Volcanic hazards: risk perception and preparedness.* New Zealand Journal of Psychology, 2000. **29**(2): p. 86-91.

[9] Knocke, E.T. and K.N. Kolivras, *Flash Flood Awareness in Southwest Virginia.* Risk Analysis, 2007. **27**(1): p. 155-169.

[10] McGee, T.K. and S. Russell, *"Its just a natural way of life..." an investigation of wildfire preparedness in rural Australia.* Environmental Hazards, 2003. **2003**: p. 1-12.

[11] McClure, P.J., *Guidelines for encouraging householders' preparation for earthquakes in New Zealand - report for Building Research,.* 2006, Victoria University of Wellington: Wellington, New Zealand.

[12] Fischhoff, B., *Risk Perception and Communication Unplugged: Twenty Years of Process.* Risk Analysis, 1995. **15**(2): p. 137-145.

[13] Jaeger, C.C., et al., *Risk, Uncertainty and Rational Action.* 2001, London: Earthscan Publications.

[14] Tversky, A. and D. Kahneman, *Availability: A heuristic for judging frequency and probability.* Cognitive Psychology, 1973. **5**: p. 207-232.

[15] Gigerenzer, G. and D. Goldstein, *Reasoning the fast and frugal way: Models of bounded rationality.* Psychological Review, 1996. **103**: p. 650-669.

[16] Slovic, P., B. Fischhoff, and S. Lichtenstein, *Perceived risk: psychological factors and social implications.* Proceedings of the Royal Society of London:Series A, Mathematical and Physical Sciences., 1981. **376**: p. 17-34.

[17] Slovic, P., et al., *Rational Actors or Rational Fools? Implications of the Affect Heuristic for Behavioral Economics*, in *Second Annual Symposium on the Foundation of the Behavioral Sciences* 2002: Great Barrington, Massachusetts, July 19-21, 2002.

[18] Zajonc, R.B., *Feeling and thinking: Preferences need no inferences. .* American Psychologist, 1980. **35**(2): p. 151-175.

[19] Finucane, M., et al., *The affect heuristic in judgements of risks and benefits.* Journal of Behavioral Decision-Making, 2000. **13**: p. 1-17.

[20] Fishbein, M. and J. Cappella, *The role of theory in developing effective health communications.* Journal of communication, 2006. **56**: p. S1-S17.

[21] Kahneman, D. and A. Tversky, *Prospect theory: An analysis of decision under risk.* Econometrica, 1979. **47**: p. 263-291.

[22] Harries, T., *Feeling secure or being secure? Why it can seem better not to protect yourself against a natural hazard.* Health, Risk & Society, 2008. **10**(5): p. 479-490.

[23] Garland, P., *Flooding – A Personal Perspective*, in *Flood Repair Network Workshop no 4 08/05/08* 2008: Embassy House, Birmingham UK.

[24] Smith, B., *Coping as a predictor of outcomes following the 1993 Midwest flood.* Journal of Social Behavior and Personality, 1996. **11**(2): p. 225-239.

[25] Bhatti, A., *Risk Perception, Culture and Communication: A South Asian Experience*, in *5th Conference of the Human Reaction and Risk Perception to Catastrophic Events: A Psychological and Cultural Perspective (Disaster and Social Crisis Research Network. Session 6); August 28-September 1, 2001*. 2001: Helsinki.

[26] Plapp, T. and U. Werner, *Understanding risk perception from natural hazards: examples from Germany*, in *RISK 21 - Coping with Risks due to Natural Hazards in the 21st Century*, D.V. Amman, Editor. 2006, Taylor & Francis Group: London. p. 101-108.

[27] Lave, T.R. and L.B. Lave, *Public perception of the Risks of Floods: Implications for Communication.* Risk Analysis, 1991. **11**(2): p. 255-267.

[28] Tapsell, S.M. and S.M. Tunstall, *I wish I'd never heard of Banbury: The relationship between 'place' and the health impacts from flooding.* Health & Place, 2008. **14**(2): p. 133-154.

[29] Lamond, J.E. and D.G. Proverbs, *Flood Insurance in the UK - a survey of the experience of floodplain residents*, in *FRIAR (1st international conference on Flood Recovery, Innovation And Response) 2-3 July 2008 One Great George Street, London, UK*, D. Proverbs, C. Brebbia, and E. Penning-Rowsell, Editors. 2008, WIT Press, Southampton: London. p. 325-334.

[30] Siegrist, M. and H. Gutscher, *Natural hazards and motivation for mitigation behaviour: people cannot predict the Affect evoked by a severe flood.* Risk Analysis, 2008. **28**(3): p. 771-778.

[31] Rotter, J., *Generalized expectancies for internal versus external control of reinforcement.* Psychological monographs, 1966. **33**(1): p. 300-303.

[32] Coleman, M. and T. DeLeire, *An economic model of locus of control and the human capital investment decision.* J of Human Resources, 2003. **38**(3): p. 701-721.

[33] McClure, P.J., F.H. Walkey, and M. Allen, *When earthquake damage is seen as preventable: attributions, locus of control and attitudes to risk.* International Journal of Applied Psychology: an international review, 1999. **48**(2): p. 239-256.

[34] Stajkovic, A.D. and F. Luthans, *Social Cognitive Theory and Self Efficacy: Going Beyond Traditional Motivational and Behavioral Approaches.* Organizational Dynamics - Field Report, 1998. **Spring 1998**.

[35] Blennow, K. and J. Persson, *Climate change: motivation for taking measures to adapt.* Global Environmental Change, 2009. **19**: p. 100-104.

[36] Cutter, S.L., et al., *A place-based model for understanding community resilience to natural disasters.* Global Environmental Change, 2008. **18**: p. 598-606.

[37] Mileti, D.S., *Factors related to flood warning response*, in *US-Italy research workshop on the Hydrometeorology, Impacts, and Management of Extreme Floods; November 1995*. 1995: Perugia (Italy).

[38] Sattler, D.N., C.F. Kaiser, and J.B. Hittner, *Disaster Preparedness: Relationships Among Prior Experience, Personal Characteristics, and*

Distress. Journal of Applied Social Psychology,, 2000. **30**(7): p. 1396-1420.

[39] Bamberg, S. and G. Moser, *Twenty years after Hines, Hungerford and Tomera: a new meta-analysis of psycho-social determinants of pro-environmental behaviour.* Journal of Environmental Psychology, 2007. **27**(2007): p. 14-25.

[40] Grothmann, T. and A. Patt, *Adaptive capacity and human cognition: the process of individual adaptation to climate change.* Global Environmental Change, 2005. **15**: p. 199-213.

[41] Grothmann, T. and F. Reusswig, *People at Risk of Flooding: Why Some Residents Take Precautionary Action While Others Do Not.* Natural Hazards, 2006. **38**: p. 101-120.

[42] Paton, D., et al., *Making sense of natural hazard mitigation: Personal, social and cultural influences.* Environmental Hazards, 2010. **9**: p. 183-196.

[43] Paton, D., et al., *Managing tsunami risk in coastal communities: Identifying predictors of preparedness.* The Australian Journal of Emergency Management, 2008. **23** (1).

A model for simulating event scenarios and estimating expected economic losses for residential buildings: preliminary results

F. Luino[1], G. Nigrelli[1], M. Chiarle[1], M. Biddoccu[2] & C. G. Cirio[3]
[1]*Consiglio Nazionale delle Ricerche (CNR),*
Istituto di Ricerca per la Protezione Idrogeologica (IRPI), Torino, Italy
[2]*Consiglio Nazionale delle Ricerche (CNR) – IMAMOTER, Torino, Italy*
[3]*Geologist, Torino, Italy*

Abstract

Floods are very common in Italy. Annually floods occur in different parts of the peninsula and cause severe damage and sometimes result in injuries or fatalities. In the period 1980–2000 the Italian State set aside 7,4 million euro for flood damage, or roughly one million euro per day.

Researching historical floods gives evidence of the frequency, magnitude, impacts, areas involved and types of damages. In spite of this wealth of information, town-and-country planning rarely takes into account these lessons of the past.

With this study the CNR-IRPI of Torino (Research Institute for Geo-Hydrological Protection) developed a flood damage estimation model for simulating event scenarios and estimating expected economic losses. Potential economic loss assessment implies knowledge of the event, exposed asset values and the degree of damage. Following a widely shared simplifying assumption, flood water level was taken as the only factor indicating event magnitude.

The model incorporates the following steps:

a) event description: definition of flood parameters (inundated area and water level), utilizing real-time measurements or data from event simulation with a hydraulic model;
b) asset damage and identification of the affected population;
c) evaluation of the damage degree as a function of event magnitude;
d) attribution of an economic value to different exposed assets;

e) quantification of economic loss by multiplying economic losses and damage severity.

The method could be used either as a forecasting tool to define event scenarios or for "real-time" damage assessment just after a catastrophe. This pilot effort can be useful for land administration bodies (land use planning, civil protection, and risk mitigation) and insurance companies in demonstrating the benefit of systematic acquisition and management of flood-related damage data.

Keywords: flood, damage evaluation, stage-damage curves, Northern Italy.

1 Introduction

Floods are causing relevant damage year after year all over the world. The average number of people who have suffered from flood damage during the period 1973-1997 amounts to more than 66 million a year. This makes flooding the worst of all natural disaster causes, among all endogenous and exogenous phenomena. The average number of flood victims is increasing: for the five-year period from 1973 through to 1977 was 19 million and has escalated sharply to 111 million for the 1988 through 1992; still further to 131 million for the 1993 to 1997 five-year period. The average death toll per year has been recorded as approximately 7,000 people for the last 25 years [1].

In the last 20 years severe hydrological events have struck Northern Italy at least eight times, taking high tolls in damage and human lives. Despite extensive flood management efforts, flood damage is increasing owing mainly to augmented human pressure on the environment, with a reduction in natural areas around watercourses.

In Italy, more than 700 agencies deal in some way with the economics of floods. The various agencies have an ordinary management, but after a disaster they can operate extraordinary competences. For all these agencies flood damage evaluation is highly important: rapid assessment of damage has to be done before repair interventions can be started. Usually, damage evaluation is carried out by expert visual inspection, but collecting comprehensive damage information in an affected area also takes time. A longer evaluation will sometimes produce very different results from those obtained in a short evaluation because of limited inspection items.

2 Conceptual framework

In 2004 the European Community initiated the "DAMAGE" project [2] in response to the need of European civil protection services to have a common methodology for the evaluation of the damage produced by a disaster. One of the project's goals was to provide public administrations (e.g. municipal. provincial and regional governments) with a management tool for damage information. For this purpose a GIS-based model was created that can simulate flood events and evaluate subsequent economic losses.

Damage caused by floods is direct and indirect. Deaths of people and animals, damage to houses and properties, agricultural crops, damage to infrastructures may be the result of direct consequence of floods. Others like, change in ecosystem or spreading of diseases or traumatic injuries defined “syndrome of the survivor” may be some of the indirect damages due to floods.

The traditional approach to damage assessment consists of detailed field survey to establish the actual loss. In the last decades attempts have been made to define methodologies for rapid assessment of damage in the aftermath of a disaster, while the main objective of this research was to devise a method that could enable real-time assessment of potential economic direct loss due to a catastrophe. The method is based on a thorough knowledge of the local area coupled with a description of some physical features of the natural event.

Given a catastrophic natural event of given intensity affecting a given area, the worth of economic losses connected to direct damages to goods depends on the number and the economic value of the units of each element in the area and on the degree of damage to the exposed units (varying from 0 = undamaged to 1 = completely destroyed).

The economic worth of loss can be defined as:

DIRECT ECONOMIC LOSS = Σi (UNIT VALUEi * N° units EXPOSED ELEMENTi * DAMAGE DEGREEi [%])
DAMAGE = State of a building or system after a disaster [3]
LOSS = Effect and impact of damage [3]
EXPOSED ELEMENTS = Population, property, economic activity, public services, environmental goods and cultural heritage items in a location exposed to danger.
DAMAGE DEGREE = VULNERABILITY = Degree of loss of one or more elements at risk resulting from a dangerous phenomenon with a given intensity and duration. This is a function of the intensity and typology of element at risk.

3 Description of the methodological model and its application to a case study

The methodological model developed for flood damage estimation can be summarized in the following steps (Figure 1):

1) Event description: definition of flood parameters (inundated area and water level). This definition will be enabled by real-time measurements or by event simulation with a hydraulic model.
2) Damaged assets identification as a consequence of the flooded area.
3) Evaluation of the degree of damage to the exposed elements as a function of event magnitude identified from measurement of floodwater depths of an event.
4) Attribution of an economic value to exposed assets.
5) Quantification of economic loss by multiplying economic value of damaged assets and degree of damage.

The method can be used to estimate the damage from the impact of floodwaters on exposed elements (direct damage) and to quantity the resulting economic loss (tangible damage). Indirect and intangible damage assessment is beyond the scope of this model.

During the development of the model, a case study was set up to test and calibrate the methodology. The Boesio Stream basin (Varese Province, Lombardy. Italy) (Figure 2) was chosen as the study sample because of its relatively small area (45 km^2), intense human pressure, availability of territorial data and historical information about the damages from past catastrophic events,

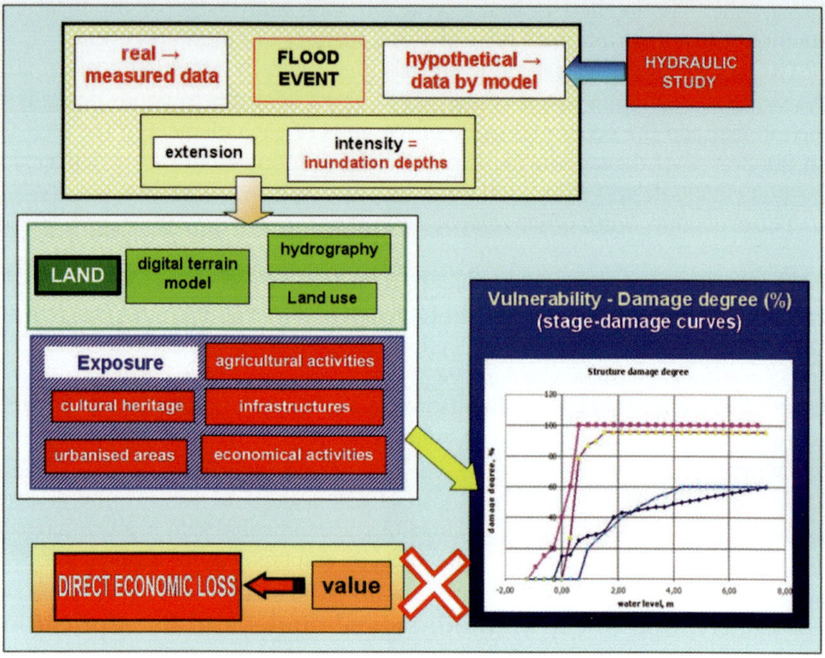

Figure 1: Conceptual model for flood damage assessment.

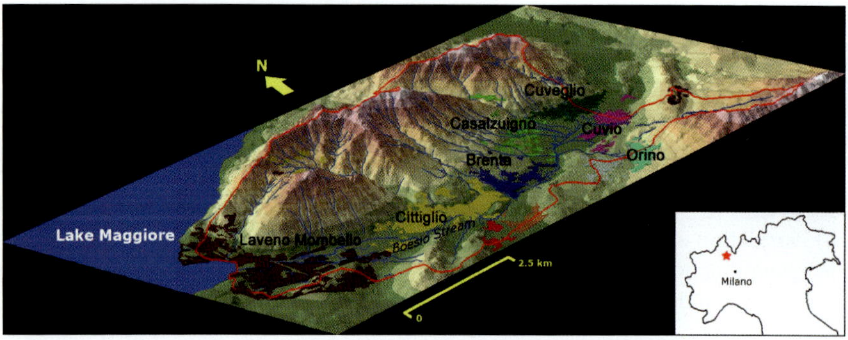

Figure 2: Location and limits of the Boesio catchment.

a recent severe flooding event for which information about damage was well documented. The method was applied to the case study to estimate flood damage to structures. Damage to agriculture and mobile goods was not taken into account because that would have required data not available for the area and additional time for research and data processing.

3.1 Geodatabase

Application of the methodology is bound by data availability for a study area; these data should be organised into a GIS database. The data can be collected and classed in two typologies:

- layers containing geographic information about the territory: topography, altimetry (DTM), hydrography, hydrology;
- layers pertaining to anthropic presence, exposed assets and land use.

As a rule, the more detailed the information, the greater the precision in potential loss assessment.

3.2 Typology of flood events

The method can be applied to two types of floods:

Real events: flooded area and inundation depths measured in the field. The method allows a quick estimate of the approximate damage extent during or immediately after an event ("nowcasting").

Event scenarios based on hydrological analyses. The event's characteristics are obtained through hydraulic models, and the methodology is applied for forecast purposes to predict the consequences of a given future event.

On May 3, 2002 the study area was struck by a severe rainfall event. The Boesio Stream and its tributaries overflowed, involving built-up and rural areas and causing heavy damage to houses and factories (Figure 3). This flood event was chosen as the reference event to calibrate and test the method, a back analysis was performed to verify process validity and reliability of the results. For this reason, the event was thoroughly investigated using surveys, interviews

Figure 3: The Hospital Parking of Cittiglio during the May 2002 flooding.

with local inhabitants, and data acquisition from the local administration databases. The inundation limits and the levels reached by flood waters were identified and graphically represented as GIS layers. Data about event-related damage and losses were collected and processed. For each inundated structure information was collected about: structure typology and use, ground floor surface area, building value, hydrometrical level reached by the floodwaters on the ground floor, damage to the structure and consequent economic loss.

3.3 Hydrologic-hydraulic study

Application of the methodology requires knowledge of a flooded area and of event intensity (floodwater level). The definition of flooded area and water level is organized into 4 phases.

1) Data collection. Input data of the hydrological study are: discharges or rainfalls measured in the sample catchment basin (and/or the adjoining basins) during the event; historical series of discharge or rainfall measures. The process is easier and more reliable if the discharge measures are available; otherwise, additional calculations are required to obtain rainfall-intensity-duration curves for an area under study, as was the case of the Boesio Stream basin.
2) Hydrological study. The purpose of hydrological analysis is to determine the real or hypothetical flood discharge, which should be assumed as a reference discharge for damage calculation. In the study area, a rainfall-runoff model was defined to calculate peak discharge values corresponding to given return periods.
3) Hydraulic modelling. The hydraulic model permits the determination of the hydrometrical levels of the riverbed and, in case of overflowing, the definition of flooded areas and of floodwater depths. Preliminary information required for the hydraulic model are:
    a) peak discharge values measured or determined using a rainfall-runoff model;
    b) morphological data about the riverbed and flood-prone areas: in the study area a detailed topographic map was generated and 237 valley bottom cross-sections were defined;
    c) contour conditions.

Hydraulic modelling was carried out using the calculated discharges (corresponding to a given rainfall return period): the absolute free water level was obtained as output that went into defining the floodable areas.

3.3.1 Definition of event scenarios

A detailed hydraulic analysis was performed on critical reach of the Boesio Stream in the Cittiglio territory. Two event scenarios were created corresponding to ordinary and extraordinary floods. A comparison between the floodable area generated by the model and the May 2002 flooded area shows that results fit well with the real situation.

3.4 Building vulnerability and stage-damage curves

By definition, vulnerability can be described as the degree of damage to a group of elements at risk resulting from a natural event with a given intensity. The degree of flood-induced damage on structures is determined by many factors, including water level. flow speed, quantity of suspended and floating load, contaminants in the water and flood duration. The published literature (Van der Sande [4]; Queensland Government [5]; FEMA [6]) reports that considering the depth of flooding as the only indicative factor for flood event magnitude is a widely shared simplifying assumption.

Given a particular type of exposed element, a relationship can be defined between the depth of flooding and the loss incurred as a percent of the element's total worth (damage degree) (Figure 4). Couples of values inundation depth-damage degree describe the direct effect of floodwater on different types of exposed elements. They define curves which are usually referred to as stage-damage curves (also called loss functions or vulnerability functions) and estimate the damage level as a value between 0 (no damage) and 1 (total destruction). These are crescent functions and are not necessarily continuous: multiple discontinuity curves are also used.

Each curve should be studied to describe the effect of floodwaters on a particular type of exposed element (distinguishing building construction type and dimension, road importance, etc.) and it can be used to simulate damage for potential future inundations. However, extrapolating data collected from place to place because of differences in building type and content can be problematic. For this reason, different curves should be created for different geographical areas and then applied to limited homogeneous territories.

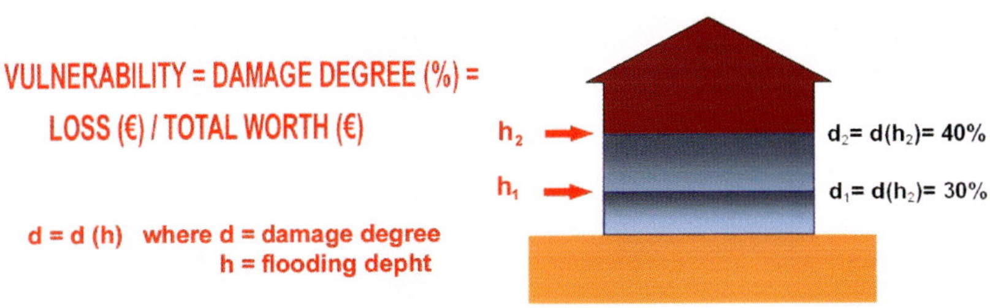

Figure 4: Given a particular type of structure, different water levels are uniquely related to different damage degrees.

The data gathered for each inundated structure during the May 2002 flood event in the Boesio Basin were employed to develop stage-damage curves.

Residential buildings with a basement constitute the bulk of the structures damaged during the May 2002 flood. For this reason, a stage-damage curve was generated only for this type of building (Figure 5). An empirical approach was followed, using the information on losses measured after the flood combined with estimates of water depth. The curve describes the effect floodwaters on the structure (masonry, floor, doors, windows, equipment, installations), without considering mobile goods.

For each building the damage degree was calculated by dividing the economic loss sustained to repair structural damage by the value of the ground floor plus basement (see section Economical evaluation of buildings).

damage degree (%) = loss (€) / [ground floor value (€) + basement value (€)]

About 100 couples of values of water level – damage degree were plotted on a dispersion diagram, and a stage-damage curve was obtained by linear interpolation. The flood depth to develop the curves includes both flooding to the basement level of the structure (negative values) and the depth of water above the ground surface (positive values).

3.5 Economic evaluation of buildings

A preliminary evaluation of the elements exposed to hydraulic risk is essential for damage quantification in either a nowcasting or a forecasting application of the model.

With regard to buildings, a punctual evaluation was applied, considering each structure in the flood-prone areas. Estimation of buildings and contents value is based on knowledge of the structural type of building and its use. For this, the study area database requires a layer containing the delimitation and the characteristics of all buildings (building use, typology, number of floors, presence or absence of basement, ground floor surface area).

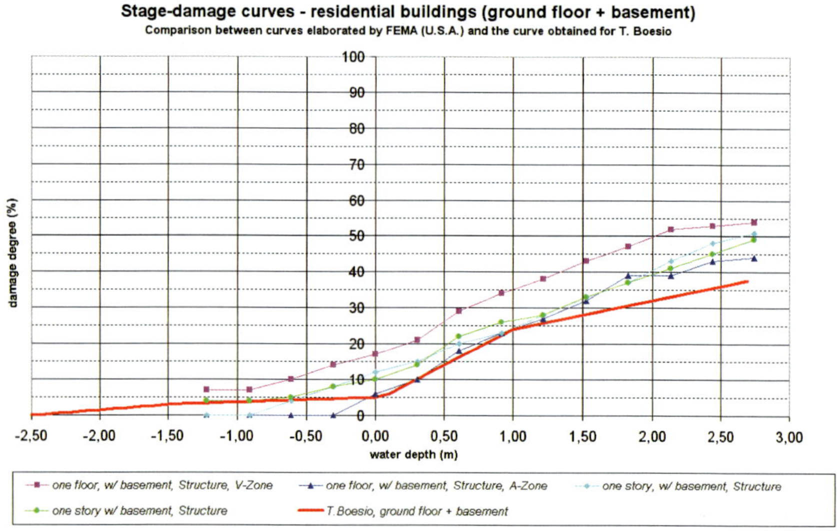

Figure 5: Stage-damage curves for residential buildings. Comparison between curves elaborated by FEMA in the U.S. (after FEMA [6], modified) and the curve obtained for the Boesio Stream.

In Italy, the unit value of buildings (€/m^2), considering only the structure and not the content, is given by the Banca dati delle quotazioni immobiliari (= Real estate and property price database) published by the Osservatorio del Mercato Immobiliare dell'Agenzia del Territorio (= District real estate market analyst services) [7]. The estimation of a building's unit value is based on its geographical location, use and typology. The database furnishes a minimum and a maximum value for each type of building and is updated every six months.
The database was used to assign a unit value to the buildings in the Boesio basin, calculating the average between the minimum and the maximum value estimated for each type of building.

In the Boesio basin, the ground floor and the basement are the only parts of a building that are vulnerable to flood damage (Figure 6). In general, Boesio floodwaters reach a maximum height of two meters above the ground. For this reason, for each building in the flood-prone area, only the value of the ground floor and the basement was considered in estimating exposure. The value of the ground floor was calculated by multiplying the ground floor surface area by the unit value. Following FEMA evaluation criteria [6], the basement value was expressed as a percent of the total structure value: for the study area the basement (cellar, boiler room, garage) value was estimated to be about 25% of the ground floor value.

3.6 Calculation of expected building damage

During the development of the methodology, a computational model for the assessment of flood expected damage was elaborated in a GIS environment. Raster information layers in grid format were employed for this purpose. Using the ArcGIS software and its Spatial Analyst extension a territorial analysis was performed by operating between different raster datasets to obtain information from superposition and combination of themes.

The computational model was applied to the case study to calculate the economic damage to residential buildings due to the May 2002 flood in Cittiglio town [8].

The following information layers stored in the database were used:

- digital terrain model;
- water surface elevation (expressed in m asl), calculated by hydraulic modelling. In this study calculated water depth values instead of those measured on field after the May 2002 flood were used because they were more uniformly distributed. Moreover, not enough floodwater levels were available for all built-up areas and buildings to perform a reliable interpolation.
- polygonal layer of built-up areas, with buildings classified by use, building mode and economic unit values in €/m2 (ground floor unit value + basement unit value).

Height H (m)	Damage (%)	Loss
-2,5	0	0
-2,0	2	€ 4.800
-1,5	3	€ 7.200
-1	4	€ 9.600
-0,5	4,5	€ 10.800
0	5	€ 12.000
+0,2	7	€ 16.800
+0,4	12	€ 28.800
+0,6	16	€ 38.400
+0,8	20	€ 48.000
+1,0	24	€ 57.600

House with basement (240,000 € of value): calculation of damage amount as a function of water height in the house

Figure 6: The loss has been evaluated as damage to real estate structure (masonry, frames, thermic and electrical systems, etc.) and not to personal property (furnishing, foodstuffs, car, etc.).

All the layers in shapefile format were convened into a grid format; the same cell size was set for each in order to obtain a one to one ratio between the cells. This bears direct matching between cells from different grids and enables operations between grid values. Considering the resolution of the data source, a cell dimension of 1 metre was chosen in the conversion phase from shapefile to raster. This cell size is suitable for representing any object or land form with sufficient detail.

The built-up area shapefile was first divided into different layers, each illustrating structures belonging to a single use category (residential, production, trade, etc.). Only the shapefile of residential buildings was afterwards used.

For all flooded locations, the ground surface elevation grid (DTM, m asl) was subtracted from the calculated water surface elevation grid (expressed in m asl), resulting in a grid that illustrates the floodwater levels above ground surface (m). Calculation of the expected economic loss was later carried out by the operations illustrated in Figure 7.

The resulting Loss grid defines the potential loss values in each cell; its attribute table contains the loss value classes and the related number of cells.

The related table of the Loss grid was exported and processed on a spreadsheet. Thus, using simple addition and multiplication, it was possible to obtain an

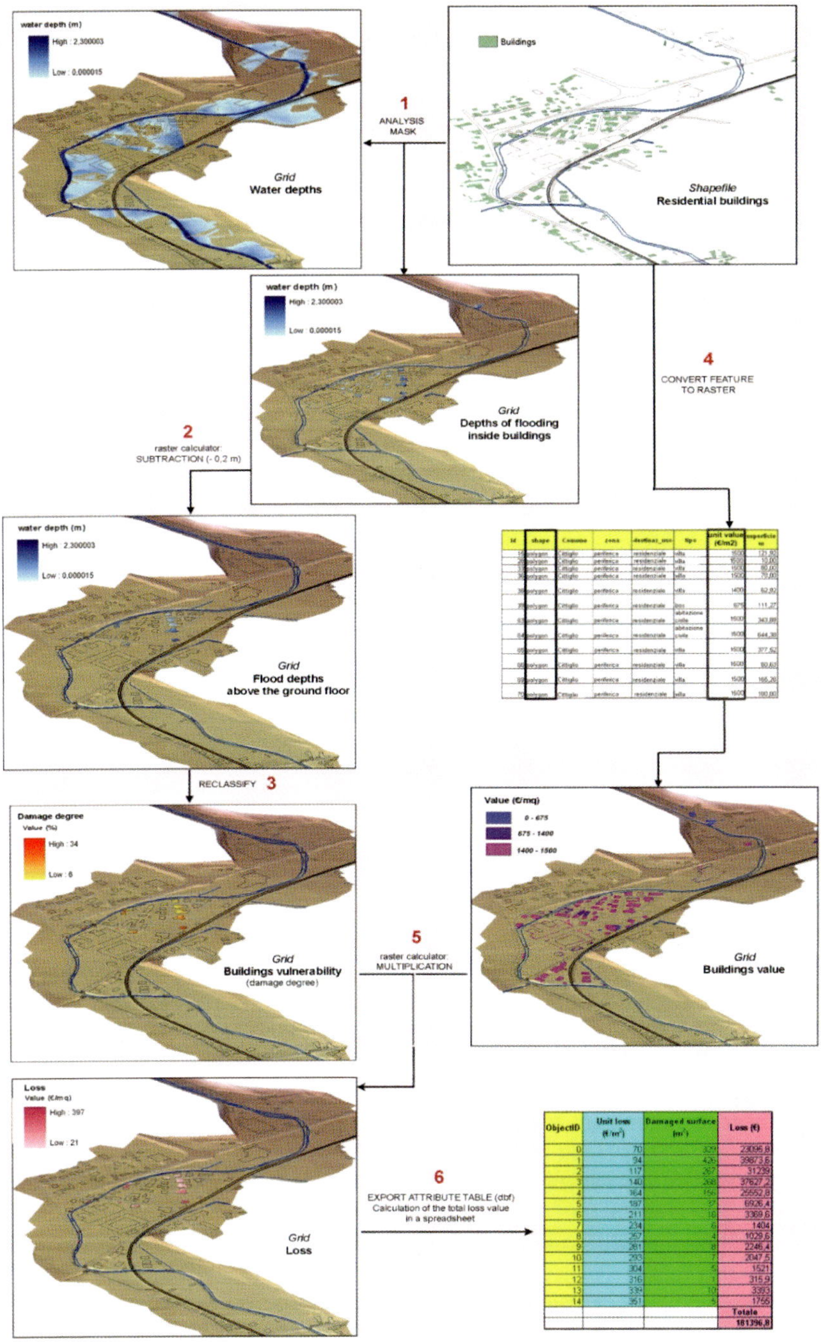

Figure 7: Calculation of the expected flood damage loss.

appraisal of the loss value for the residential buildings in Cittiglio. The calculated loss amounts to €181,000. Losses for the Boesio flood-related damage to the structure of residential buildings and officially declared by Cittiglio Municipality amounted to €110,000. Overlooking the approximation of the calculation model and the need to calibrate the method with other case studies, the difference between the real loss and the calculated value is ascribable to the damage assessment criteria. Only losses over €2500 for damage to principal residences were considered refundable and so reported, making a total loss of €110,000 certainly an undervalued sum.

4 Conclusions

The objective of effective management in the post-emergency phase is to provide public administrations with a simple and rapid instrument that can facilitate the various steps of the process, from initial damage reporting following a natural disaster to emergency interventions, to communication of expenses sustained for such interventions, to claims for disaster aid and relief, to payment for claims, taking into account the autonomy of the local bodies involved.

In recent years, a growing need has emerged at the international level for quantifying a priori flood damage with estimates based on predicted damage (potential damage) for events with an assumed probability of occurrence.

For many fields of application, the ability to quantify damage is of vital importance in the decision-making phase: from the engineering of structural interventions for risk mitigation to land-use planning, from flood insurance to civil protection. This methodology, precisely because an event has not occurred, appears useful in any area and is critical for calibrating numerical simulations to estimate flood risk. However, more detailed work is needed to collect more information.

Within the framework of this study, event simulation and calculation of flood damage was carried out and tested in the Boesio stream basin (province of Varese-northern Italy) which had been struck by flooding in May 2002. In-depth analysis of the area over the course of about one year furnished a wealth of data on the area's geographic characteristics, anthropic activities, and effects and damage stemming from past floods. During the study, there arose the need to collect more detailed data on the structures exposed to floods and the damage sustained.

Numerous problems were encountered during application of the method: a) difficulty in retrieving historical data on flood events; b) heterogeneity of information in the damage reports filed with the various different municipal offices; c) overestimation of damage in relation to payment claims submitted by private citizens; d) difficulty in collecting useful information from flood victims, even after a relatively recent event, owing to the lack of documentation and/or photographs demonstrating the damage; e) difficulty in interpreting the criteria the municipalities applied for making claims payments.

Nonetheless, the experience acquired during the study and the results achieved were highly promising in themselves and pointed to the need to initiate an assessment and retuning phase to optimize the instrument for future application.

Moreover, the methodological and technical solutions that emerged from this study open interesting prospects for extending the application to areas that require support for spatial analysis (e.g., technological risk, seismic risk, environmental impact assessment, insurance programs, etc.) and instruments for the collection and diffusion of information about an area under study.

The transversal use of GIS technologies to support the diverse information requirements for disaster management revealed the strengths and criticalities of the system's capacities, indicating steps for further developing its architecture and knowledge in managing it.

A more rational management of post-disaster damage reporting and the creation of event and damage scenarios will require usage of a national standard technical language to make damage description more objective and less subject to personal interpretation. With a view to accomplish this, standard forms should be used to indicate the type of event, the effect on topography, the structures and infrastructure damaged, a general estimate of damage (to public and private property), with information about eligibility for damage claims.

References

[1] Unesco, http://www.unesco-ihe.org

[2] DAMAGE Project (2004) – Developpement d'Actions pour Ie Marketing et la Gestion post-evenements. Programme INTERREG IIIB Medocc.

[3] GEORISK (2000) – Risk management terminology, 3 pp., http://www.georisk.com/index.aspx

[4] Van der Sande, C. (2001) – River flood damage assessment using IK.ONOS imagery, E.C. Joint Research Centre & S.A.I., 77 pp.

[5] Queensland Government (2002) – Disaster loss assessment guidelines, Illycroft Pty Ltd, 111 pp.

[6] FEMA (2003), Hazard MH – Multi-hazard Loss Estimation Methodology, Federal Emergency Management Agency.

[7] Agenzia del Territorio (2005) – Osservatorio del Mercato Immobiliare, Banca Dati delle Quotazioni Immobiliari. Agenzia del Territorio. www.agenziaterritorio.it/servizi/osservatorioimmobiliare.

[8] Luino F., Biddoccu M., Cirio C.G., Agangi A., Giulietto W., Godone F. & Nigrelli (2009) – Application of a model for the evaluation of flood damage. GeoInformatica, Springer Ed., Vol. 13, Issue 3, 339-353.

Flooding tendencies for the city of Ensenada, Baja, California, Mexico, 1948–2004

J. Soares[1], R. Blanco[2], C. García[3], E. Inzunza[1] & P. Rousseau[1]
[1]*Faculty of Engineering,*
Autonomous University of Baja California UABC in Ensenada, Mexico
[2]*Oceanologic Research Institute IIO of the Autonomous University of Baja California UABC in Ensenada, Mexico*
[3]*Independent Researcher*

Abstract

Precipitation data collected from 4 stations located in the city of Ensenada, Baja California Mexico from 1948 to 2004 were analyzed and associated with damaging floods. Precipitation values from 1948 to 1977 did not exceed 35mm. in 24 hours; however, rainfall daily reports of 51, 52 and 81mm. were found in 1978, 1979 and 1980, respectively, exceeding the 50mm. of water/day capacity of the sewage system for the city causing considerable infrastructure damage and even the loss of lives. No values greater than 46mm have been observed from 1981 to 2004 yet landslides have been frequently observed in areas with steep slopes and alluvial material in some parts of the city. Floodplain occupancy, lack of maintenance of drainage systems as well as inundations in low land areas, account for the main flooding problems for the city. A map is developed in this study showing the different flooding zones associated with the amount of precipitation for the city. Remedial options are offered for city planners
Keywords: damaging flood, drainage systems, landslides, anthropogenic impact.

1 Introduction

Archived rainfall data for the city of Ensenada, Baja California in Mexico is relatively recent (1948-) compared with stored rainfall data for our neighbouring San Diego, California (1850-), in the United States. Although Ensenada is located approximately 120 km south of the border with the U.S.A., there has not been a study that relates the amount of water precipitation in the city of San

Diego, to quantities for Ensenada which could lead to some developments in adequate weather prediction, timely flood-warning systems and management plans to prevent or reduce flooding effects that act upon the wellbeing of the community. This paper is the first of its kind and is built upon a short history of observed damaging flooding for our city to improve the understanding of the water flow behaviour as housing development increases.

In general, the study area includes zones with smooth surface features constituted by alluvial deposits of gravel, sands and clay, while higher areas are composed mainly by granitic rocks. The majority of the population is located on the alluvial soils that derive from erosion of the surrounding hills, as well as deposits from streams. Quaternary lacustrine and marshy soils occupy zones closer to the ocean. The average temperature for the city is approximately 18.1^{0}C [1] and drought periods are present for the most part of the year so no infrastructure for collecting rain has been set up.

The municipal sewage system was built to hold a maximum of 50mm of rain/day and precipitation values that surpass this limit are causing considerable damage to the city's infrastructure; apart from this, sewage is filled with trash falling from the sewer drain and sediment that has been transported along with the rain reducing the 50 mm and flooding important parts of the city (Figure 1).

Figure 1: Flooding of ISSTECALI Hospital in 1978 due to saturation of the sewage system.

In 1995 a Land Use Chart was produced by city planners projecting housing developments towards the NE part of the city, where granitic rock composes the terrain; county government change resulted in housing development towards the south of the city, invading regions for sowing where land had been cultivated for more than 20 years in the stream plains.

Seismic velocity profiles were made in 1997 [2] showing more than 30 m of layer thickness for the alluvial deposits to the South of Ensenada whereas in the northern part of the city known as the Downtown Area, thickness of the alluvial bed is approximately 5 m.

Figure 2a shows the Downtown Area of Ensenada, before land was reclaimed from the sea and occupied with tourist leisure activities and Port operations. The Downtown Area was filled with sand and alluvial deposits below sea level. When abundant rain is present the Downtown Area inevitably floods and traffic jams are unavoidable. Borehole results obtained in 1993 by the "Laboratorio de Materiales" [3] (Figure 2b) showed freatic water levels at a depth of 1.10m. This level is influenced by tidal changes as it is very close to the ocean.

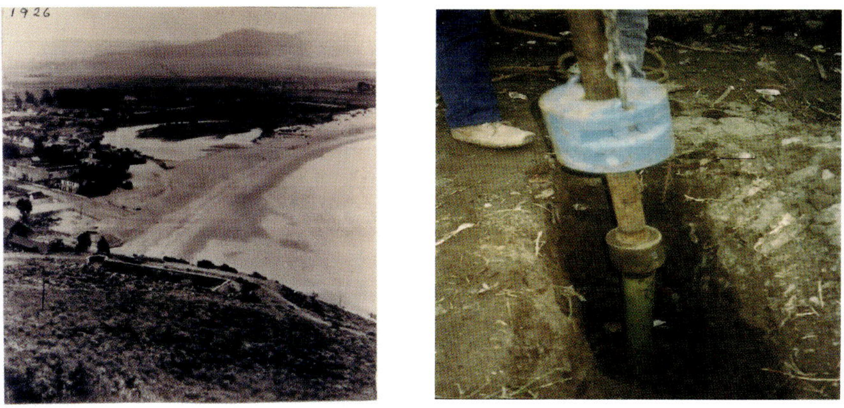

Figure 2: a) Ensenada in 1926 before land was reclaimed from the sea. b) Borehole drilling in the Downtown Area.

2 Data analyses and interpretation

Data precipitation records from four local meteorological stations over a period from 1948 to 2004 were analyzed. From the four stations chosen, the data from station Punta Banda showed extremely gappy records probably due to a poor maintenance program; data from this station proved useless for the analysis and was discarded, allowing us to use data from 3 stations only: La Presa (ELP) located at coordinates N31°53'45", 116°35'47"W, El Ciprés (ECP) at N 31°47'48", 116°35'37"W, and San Carlos (ESC) at N 31°47'03", 116° 27'00"W. An initial visual inspection of the time series for the three remainder stations showed that they have different record lengths: ELP station runs from 1948 to 2004, while ECP station has data from 1976 to 2004; ESC station started to record data from 1962 up to 2004. For the purpose of this work, time series were cut to match the same length (1976-2004) to allow fair time comparisons between them. Tables 1, 2 and 3 only show data for the months for the rainy season considering that no significant rainfall values are seen during the summer (dry season).

Table 1: Monthly accumulated precipitation (Map) in [mm] for El Ciprés (ECP) Weather Station, for the rainy season period. M is used as a symbol representing the months and ▲ is used when no data was available.

year	M	MaP [mm]	M	MaP [mm]	M	MaP [mm]	M	MaP [mm]	M	MaP [mm]	M	MaP [mm]	M	MaP [mm]
1969	Jan	▲	Feb	▲	Mar	▲	Apr	▲	Oct	▲	Nov	▲	Dec	▲
1978	√	113	√	112	√	124	√	▲	√	▲	√	▲	√	▲
1979	√	109	√	▲	√	▲	√	▲	√	▲	√	▲	√	▲
1980	√	133	√	120	√	▲	√	▲	√	▲	√	▲	√	▲
1993	√	117	√	▲	√	▲	√	▲	√	▲	√	▲	√	▲
1995	√	▲	√	▲	√	▲	√	▲	√	▲	√	▲	√	▲
1998	√	52	√	**217**	√	79.7	√	▲	√	▲	√	▲	√	▲

Table 2: Monthly accumulated precipitation (Map) in [mm] for SanCarlos(ESC) Weather Station, for the rainy season period. M is used as a symbol representing the months and ▲ is used when no data was available.

year	M	MaP [mm]	M	MaP [mm]	M	MaP [mm]	M	MaP [mm]	M	MaP [mm]	M	MaP [mm]	M	MaP [mm]
1969	Jan	171	Feb	115	Mar	▲	Apr	▲	Oct	▲	Nov	▲	Dec	▲
1978	√	202	√	157	√	169	√	▲	√	114	√	▲	√	95
1979	√	157.4	√	▲	√	138	√	▲	√	▲	√	▲	√	89
1980	√	219	√	211	√	117	√	▲	√	▲	√	▲	√	▲
1993	√	329	√	185	√	245	√	▲	√	▲	√	▲	√	▲
1995	√	147	√	▲	√	121	√	▲	√	▲	√	▲	√	▲
1998	√	▲	√	**374**	√	▲	√	▲	√	▲	√	▲	√	▲

Table 3: Monthly accumulated precipitation (Map) in [mm] for La Presa (ELP) Weather Station, for the rainy season period. M is used as a symbol representing the months and ▲ is used when no data was available.

year	M	MaP [mm]	M	MaP [mm]	M	MaP [mm]	M	MaP [mm]	M	MaP [mm]	M	MaP [mm]	M	MaP [mm]
1969	Jan	108	Feb	99	Mar	▲	Apr	▲	Oct	▲	Nov	▲	Dec	▲
1978	√	159	√	148	√	165	√	▲	√	▲	√	▲	√	86
1979	√	118	√	▲	√	126	√	▲	√	▲	√	▲	√	▲
1980	√	186	√	184	√	▲	√	▲	√	▲	√	▲	√	▲
1993	√	217	√	▲	√	▲	√	▲	√	▲	√	▲	√	▲
1995	√	135	√	124	√	▲	√	▲	√	▲	√	▲	√	▲
1998	√	▲	√	**237**	√	93	√	▲	√	▲	√	▲	√	▲

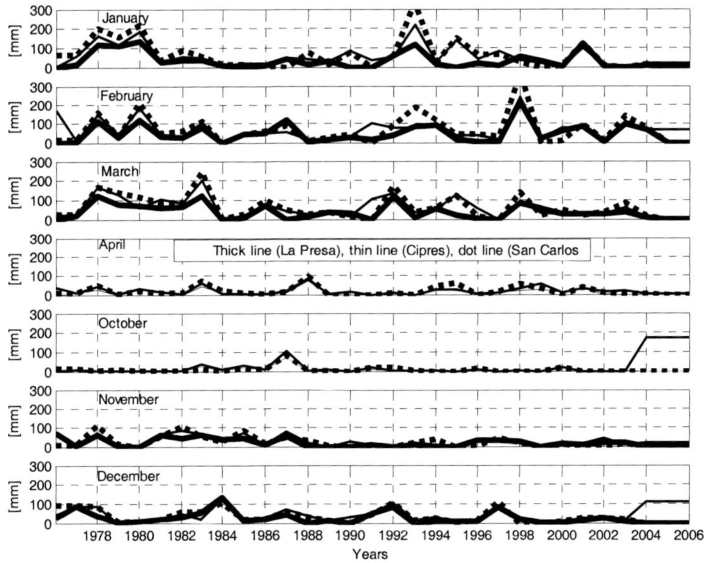

Figure 3: Overlapped total precipitation/year data for the three Stations used in this study.

Overlapped data for the three stations were graphed for a clearer comparison and are shown in Figure 3. Mainly January, February and March are known for the arrival of storms from the NW.

It is observed that important rainfall peaks differ in amplitude between the three stations but show a good agreement in time; major rainfall amplitudes (>300 mm) are registered at ESC, which is located to the SE of the city where construction companies are building housing developments while ELP station (with >200 mm of rainfall as its maximum value) is located to the north of Ensenada, at the López Zamora Dam where city planners in 1995 projected the housing developments. ECP station showed the lowest values of the group (less than 200mm). Data shows other precipitation peaks, although minor, occurring in April, suggesting the presence of occasional out-of-season showers. In early November, rainfall starts to exhibit increasing values, suggesting the prelude of the rainy season.

The most relevant rainfall peaks occur in the years 1978-1980 and 1993. Comparisons between stations clearly show that February is the most important month in terms of accumulated precipitation, followed by the month of January.

To find which part of the city receives the most amount of water during the rainy season we estimated the correlation coefficients (cc) between the stations. For stations ECP and ELP, cc=0.77; for ECP and ESC, cc=0.76 and between ECS and ELP cc=0.94. These numbers indicate that the city receives approximately the same amount of rainfall in the northern sector than in the southern sector. Similar strategies to manage the excess of rainfall along the city must be applied in the event of the presence of a severe storm.

Slope analysis using a digital elevation model with a 30 × 30 m grid [4] along with field corroborations was done to properly denote flooding areas and their main causes as a first attempt for preparing a zoning of the city.

Fieldwork gave a good basis for classifying the main factors contributing to the flooding problems:

a) Inadequate drainage systems.- Massive housing expansion towards the south of the city are built over stream plains (Figures 4a), 4b)) and the major construction companies did not consider a rain sewer system, leaving the domestic drainage system as the sole means of conducing the rain; this system is already filled with waste and sediments and eventually leads to overflowing out of the gutters to the streets wasting the much needed rain water.

Figure 4: a) Major Housing Development to the South of Ensenada built over a Stream Plain. b) Anthropogenic impact inducing flooding as incorrect deviation of runoff water concentrated from developed areas saturates the soil.

Figure 5: a) Landslide in the SW of Ensenada where hills are unprotected by recent wildfires. b) Saturation of the Soil in a Housing Complex south of the city where sowing was the major activity before the county government changed the use of the soil.

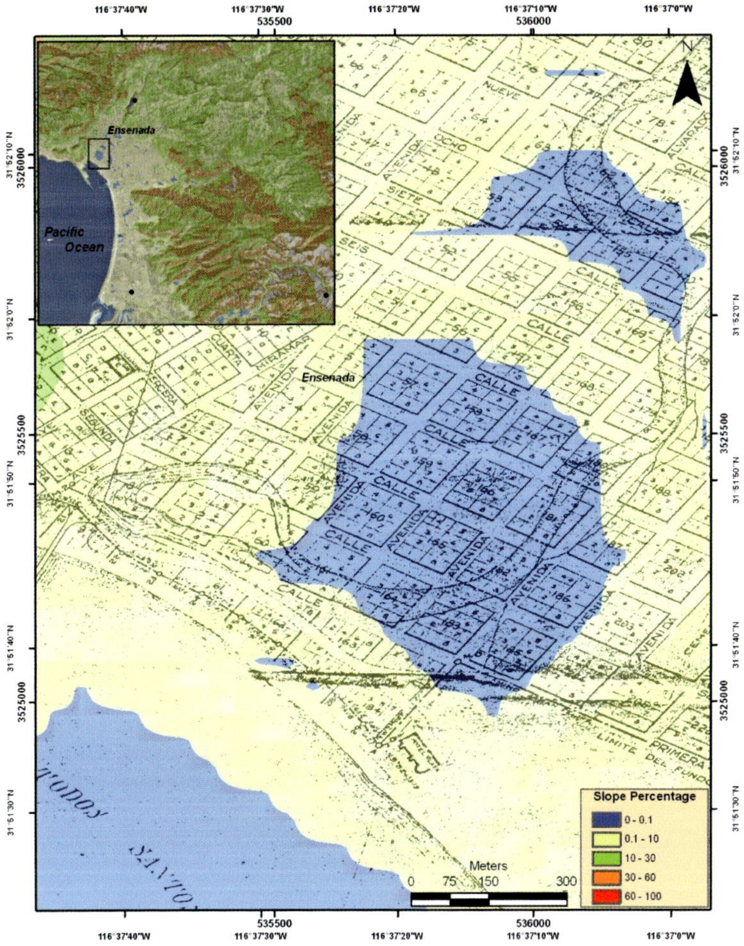

Figure 6: Map showing flooding zones for the Downtown Area of Ensenada. Far left: Black dots indicate the location of the Weather Stations used in this study.

b) Landslides- Landslides are generated by two major factors: Loss of vegetation and water saturation. Loss of vegetation is caused by wildfires induced by hot Santa Ana Winds occurring in late fall resulting in destabilization of surface soil and an increase in the velocity of surface water runoff causing mudflows in lower areas where clays form the hill areas (Figure 5a)). South of the city, landslides occur due to water saturation of the soil for the housing developments built over stream plains (Figure 5b))

c) Terrain Morphology- The city has 3 main streams: El Gallo, Ensenada and El Campillo. Although deviated by population growth, traces of their original routes remain in the form of small watersheds along the main streets, converting into ponds when it rains where cars can get stuck. Since the year 2000, great effort has been done by city managers in covering the streets with concrete pavement, but the original drainage system has not been updated to fit the requirements of an increasing population resulting in water running over the pavement raising the flow and volume which worsens the flooding. Currently the city is building 3 bridges across the Ensenada Stream in 3 of the main streets in the Downtown Area to release traffic jams when it rains.

A map was developed for this study, showing the slope percentages for the city (Figure 6). Regions below sea level and with slope percentages of 0 – 0.1 are considered susceptible to flooding.

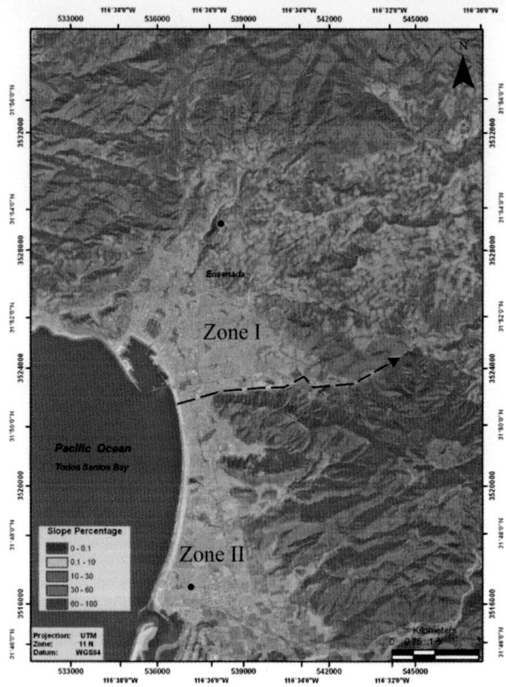

Figure 7: Macrozonation for the city of Ensenada according to the main flooding causes. The arrow points to the site where Arroyo El Gallo is originated.

Considering all the factors that contribute to the flooding of Ensenada, we decided to make a macrozonation for the city, as a first attempt to characterize the major problems that afflict the region. Figure 7 shows the consensus division made by the authors in this study. Zone I flooding occurs mainly because of the problems arising by the land reclaimed from the sea which is below sea level. Zone II flooding occurs mainly because the soil is not fitted for large housing development and deviation of the stream channel into smaller dug passages with no concrete for bottom lining and slope revetment. Both Zones have inadequate drainage systems.

3 Conclusions and recommendations

Due to its hydro-climatic and geologic characteristics, Ensenada produces large quantities of sediment deposits. Based on observations, it is imminent to execute Works to trap the sediments at the sub-basin level and to increase the maintenance of the drainage system. Continuous maintenance programs must begin in the month of October and continue to April, as those are the months expected to have the heaviest rainfalls. As it is financially possible for the city, it is imperative to change the drainage system to a larger one that serves the population necessities. At all costs, city planners should avoid housing developments over areas of stream plains. It is also important to assemble a separate system for collecting rain in Zone 2. Although its topography is more uneven, similar actions have been proposed for Tijuana, a city located 100 km. North of Ensenada by the Subdirector of the State Commission for the Public Services of Tijuana (CESPT) [4] who states that the one of the main problem of the drainage system is the garbage blocking the system. No action has been taken by city planners.

Reforestation in higher areas affected by wildfires is also recommended.

References

[1] Portal Ciudadano de Baja California, México. *Ensenada.* http://www.bajacalifornia.gob.mx/portal/nuestro_estado/municipios/ensenada/clima.jsp

[2] Soares J. (2003) Aplicación de la microzonación sísmica a la vulnerabilidad de estructuras críticas en la ciudad de Ensenada, B. C. Tesis de Maestría. Centro de Investigación Científica y de Educación Superior de Ensenada.

[3] Laboratorio de Materiales.(1993). *Estudio de Mecánica de Suelos, Sección Primera. Ensenada, Baja California.* Reporte Técnico para Ginsa Construcciones, S.A. de C.V.

[4] San Luis Adolfo Noticias Frontera.info. *Es insuficiente el drenaje pluvial.* Retrieved June 10, 2009 from http://www.arq.com.mx/noticias/Cached_Pages/3186.html

Landslides

*Edited by: **S. MAMBRETTI**, Politecnico Di Milano, Italy*

This volume is the second in the new Safety and Security Engineering series that is designed to provide a comprehensive view on risk mitigation. The book is devoted to landslides and debris flow, addressing the need for a better understanding of these increasingly frequent phenomena. With better understanding comes a greater ability to manage the attendant risk.

Landslides contains selected research papers presented at Wessex Institute of Technology Conferences. The Authors have revised their papers to bring them up to date and to integrate them into a coherent volume on the topic. The book includes the following chapters: Ranging scales in spatial landslide hazard and risk analysis; From national landslide database to national hazard assessment; An aid in the most accurate rainfall thresholds evaluation; Computer analysis of slope failure and landslide processes caused by water; The management of territorial risks: which integration between planning instruments, evaluation tools, and certification procedures; Analysis of the stability of slopes reinforced by roots; Experience with treatment of road structure landslides by innovative methods of deep drainage; Strategic program for landslide disaster risk reduction: a lesson learned from Central Java, Indonesia; Erosion of forestry land: causes and rehabilitation; Landslide in a catchment area of a torrent and the consequences for the technical mitigation concept; Slope instability along some sectors of the road to La Bufadora.

The book will be a valuable reference for professionals, scientists, and managers concerned with prediction and management of the risk of landslides and debris flows.

Series: Safety & Security Engineering
ISBN: 978-1-84564-650-9 eISBN: 978-1-84564-651-6
Forthcoming 2012 / 144pp / apx £64.00

WIT*Press*
Ashurst Lodge, Ashurst, Southampton,
SO40 7AA, UK.
Tel: 44 (0) 238 029 3223
Fax: 44 (0) 238 029 2853
E-Mail: witpress@witpress.com

Disaster Management and the Human Health Risk II

Reducing Risk, Improving Outcomes

Edited by: ***C.A. BREBBIA****, Wessex Institute of Technology, UK;* ***A.J. KASSAB****, University of Central Florida, USA and* ***E.A. DIVO****, Daytona State College, USA*

This book contains papers presented at the second International Conference on Disaster Management, held in 2011 in Orlando, Florida. Florida is no stranger to the increasing number of natural disasters affecting millions of people around the world, periodically suffering the ravages of hurricanes, with a recent disastrous sequence of four hurricanes having delivered devastating blows to the State from both Gulf and Atlantic coasts in 2004. Most recently, Florida's economy was seriously affected by the Deep Horizon oil spill in the Gulf of Mexico in spite of its coastline having been mostly spared from any major damage.

The 2011 Disaster Management Conference attracted outstanding contributions from researchers throughout the world. The collected papers, published in this book, reflect the excellent work of all contributing authors and the care taken by the Scientific Advisory Committee and other colleagues in reviewing the presentations. Topics covered include: Disaster Analysis, Monitoring and Mitigation; Emergency Preparedness; Risk Mitigation; Surveillance and Early Warning Systems; Socio-Economic Issues.

WIT Transactions on The Built Environment, Vol 119
ISBN: 978-1-84564-536-6 eISBN: 978-1-84564-537-3
Published 2011 / 336pp / £145.00